31st Annual Saudi-Japan Symposium on Technology in Petroleum Refining and Petrochemicals 2022

Dhahran, Saudi Arabia
12-13 December, 2022

ISBN: 978-1-7138-7084-5

Printed from e-media with permission by:

Curran Associates, Inc.
57 Morehouse Lane
Red Hook, NY 12571

Some format issues inherent in the e-media version may also appear in this print version.

Printed with permission by Curran Associates, Inc. (2023)

For permission requests, please contact King Fahd University of Petroleum & Minerals at the address below.

King Fahd University of Petroleum & Minerals
Dhahran 31261
Kingdom of Saudi Arabia

Phone: +966 (13) 860 0000

info@kfupm.edu.sa

Additional copies of this publication are available from:

Curran Associates, Inc.
57 Morehouse Lane
Red Hook, NY 12571 USA
Phone: 845-758-0400
Fax: 845-758-2633
Email: curran@proceedings.com
Web: www.proceedings.com

الندوة السنوية السعودية اليابانية الحادية والثلاثون

31st Annual Saudi-Japan Symposium

التقنية في الوقود والبتروكيماويات

Technology in Fuels & Petrochemicals

Proceedings

19-18 جمادى الأولى 1444هـ
December 12-13, 2022

KFUPM Research Institute
King Fahd University of Petroleum & Minerals
Dhahran, Saudi Arabia

Symposium supported by: KIKX

Preface

This is the Proceedings of the 31st annual Saudi Japan symposium "*Technology in Fuels & Petrochemicals*" held on December 12-13, 2022. This event, jointly organized by KFUPM, Saudi Aramco, the Japan Petroleum Institute (JPI), and Japan Cooperation Center, Petroleum (JCCP), brings together researchers from industry and academia to discuss recent downstream technologies related to refining and petrochemicals. The main theme of the Symposium is *"Innovative Catalyst Development"*.

At this year's symposium, there are 21 presentations featuring 21 papers and 10 posters. The topics include carbon neutrality, innovative catalysis, hydrogen & chemicals production, and waste & CO_2 utilization. Distinguished panel of speakers come from Aramco R&DC, KFUPM research centers/academic departments, Intl. Center for Synchrotron Radiation, ENEOS, Japan Advanced Institute of Science & Tech., and Universities of Tokyo, Shizuoka and Tohoku.

We extend our thanks to KFUPM, KIKX, Saudi Aramco, JPI, JCCP, distinguished speakers and attendees for their active participation and look forward to your participation in our next symposium.

Dr. Hassan S. Alasiri
Director, Center for Refining & Advanced Chemicals
http://ri.kfupm.edu.sa/crp/
KFUPM-Dhahran, Saudi Arabia

Contents

****

31st Annual Saudi-Japan Symposium – 2022

Technology in Fuels & Petrochemicals

Innovative Catalyst Development

Venue: KFUPM Dhahran Techno Valley, Innovation Cluster Bldg., Auditorium 1-111
December 12-13, 2022

Day One: Monday, December 12, 2022

OPENING REMARKS SESSION — ***Chairman: Dr. Hassan Al-Asiri, KFUPM***

8:30 Opening Remarks

- ***Dr. Muhammad M. Al-Suggaf,*** *President, KFUPM*
- ***Mr. Makoto Yokota,*** *Managing Director Secretary General, JCCP*
- ***Mr. Ahmad O. Al-Khowaiter,*** *CTO, Saudi Aramco*
- ***Dr. Atsushi Muramatsu,*** *President, The Japan Petroleum Institute, JPI*

SESSION ONE CARBON NEUTRALITY AND CIRCULAR ECONOMY — ***Chairman: Prof. Keiichi Tomishige, Tohoku University***

9:00	**1.**	**Keynote: TBD,** *Dr. Gerard de Nazelle, Saudi Aramco R&DC*
9:30	**2.**	**Keynote: JPI direction along carbon neutrality and introduction of our research trend for use of NanoTerasu, next generation synchrotron radiation facility,** *Dr. Atsushi Muramatsu, President of JPI, Associate Executive Vice President and Director of International Center for Synchrotron Radiation Innovation Smart, Japan*
10:00		***Coffee break***

SESSION TWO INNOVATIVE CATALYSTS, UPGRADING PROCESSES — ***Chairman: Dr. Khalid Almajnouni, Saudi Aramco***

10:20	**3.**	**Residue upgrading by hydroprocessing deasphalted oil** *Dr. Koichi Matsushita, ENEOS Corporation, Japan*
10:40	**4.**	**A high-throughput approach to study cation-exchanged zeolites** *Dr. Hassan Aljama, Saudi Aramco R&DC*
11:00	**5.**	**Catalytic manufacture of ethylene–1-hexene thermoplastic elastomer: Thermodynamic process development,** *Dr. Muhammed Atiqullah, Center Refining & Adv. Chemicals, KFUPM*
11:20	**6.**	**Hierarchical nano-sized zeolite Beta development for crude oil hydrocracking catalysts** *Dr. Lianhui Ding, Saudi Aramco R&DC*
11:40		***Prayer & Lunch Break***

Each presentation includes 5-minutes Q&A Rev. Nov. 28, 2022

Symposium supported by: KIKX

31st Annual Saudi-Japan Symposium – 2022

aramco

Technology in Fuels & Petrochemicals

Innovative Catalyst Development

Day One: Monday, December 12, 2022

POSTER SESSION		*Chairman: Dr. Rajesh* Theravalappil, *KFUPM*
13:00	P1.	**Chemometrics study of sulfur reactivity in hydrodesulfurization process** *Ms. Hutoon Alhargan, Saudi Aramco R&DC*
13:05	P2.	**Development of hierarchical CoMo/Beta zeolite for BTX production** *Ms. Eman Alhaher, Saudi Aramco R&DC*
13:10	P3.	**Development of new mesoporous materials for refining processes** *Dr. Sathiyamoorthy Murugesan, Center Refining & Adv. Chemicals, KFUPM*
13:15	P4.	**Thermal liquefaction of waste polystyrene plastic to liquid hydrocarbon fuel using ethanol as a solvent**, *Dr. Nabeel Ahmad, Center Refining & Adv. Chemicals, KFUPM*
13:20	P5.	**Oxidative dehydrogenation of n-butane to produce petrochemical key building blocks (ethylene, propylene, and butadiene)**, *Mr. Ahmad Sabban, CHE, KFUPM*
13:25	P6.	**Controlled autoxidation of heavy petroleum to produce carbon fiber precursors** *Mr. Lahmady S. Mohamed, CHE, KFUPM*
13:30	P7.	**Characterization of soot from atmospheric combustion of diesel/gasoline-oxygenate blends**, *Mr. Mohammed Qasem, CHE, KFUPM*
13:35	P8.	**Turning municipal solid waste into fuel via integrated thermochemical and electrochemical processes**, *Mr. Achmad Putra, CHE, KFUPM*
13:40	P9.	**CO_2-assisted oxidative dehydrogenation of propane to propylene over MoO_3 catalyst** *Mr. Suleiman Magaji, CHE, KFUPM*
13:45	P10.	**Effects of metal support interaction on dry reforming of methane over Ni/Ce-Al_2O_3 catalysts**, *Mr. Ariel Gursida, CHE, KFUPM*
13:50		***Poster Session Ends***

SESSION THREE	ARTIFICIAL INTELLIGENCE, DIGITALIZATION	*Chairman: Dr. Jihad Badra, Saudi Aramco*
14:00	**7.**	**Digital chemistry guided development of novel ligands for selective ethylene tetramerization**, *Dr. Mohamed Elanany, Saudi Aramco R&DC*
14:20	**8.**	**High-throughput screening and literature data driven machine learning (ML) assisting discovery of La_2O_3-based catalysts for low temperature oxidative coupling of methane,** *Associate Prof. Shun Nishimura, Japan Advanced Institute of Science and Technology, Japan*
14:40	**9.**	**Optimal design of AI-based models integrated with Ensemble ML paradigms for crude to chemicals applications**, *Dr. Abdulkadir Tanimu, IRC Refining & Adv. Chemicals, KFUPM*
15:00		***Day One Ends***

31st Annual Saudi-Japan Symposium – 2022

aramco

Technology in Fuels & Petrochemicals

Innovative Catalyst Development

Day Two: Tuesday, December 13, 2022

SESSION FOUR   HYDROGEN PRODUCTION   ***Chairman: Dr. Nabil Al Yassir Honeywell UOP***

8:30 10. Disruptive water electrolyzer system for cost-effective hydrogen production: Electrolyte engineering, *Prof. Kazuhiro Takanabe, Dept. Chemical System Engineering, University of Tokyo*

8:50 11. Development of low temperature steam reforming catalyst for hydrogen production for fuel cells, *Dr. Shakeel Ahmed, IRC Refining & Adv. Chemicals, KFUPM*

9:10 12. Thermo-catalytic production of hydrogen from H_2S: challenges and opportunities
Dr. Zainab A. Aithan, Saudi Aramco R&DC

9:30 13. Titania supported bimetallic catalyst for hydrogen production from methane
Dr. Wasim Ullah Khan, IRC Refining & Adv. Chemicals, KFUPM

9:50 14. Ammonia to hydrogen: cracking technology, *Dr. Stephen Paglieri, Saudi Aramco R&DC*

10:10 Coffee Break

SESSION FIVE   CHEMICAL PRODUCTION   ***Chairman: Prof. Takanabe, The University of Tokyo***

10:30 15. *Redox dehydrogenation of propane utilizing lattice S^{2-} in metal sulfide catalyst*
Associate Prof. Ryo Watanabe, Shizuoka University, Japan

10:50 16. An engineered bimetallic Fe-Cu-MOR zeolite catalyst for direct oxidation of methane to methanol *Dr. Ijaz Hussain, IRC Refining & Adv. Chemicals, KFUPM*

11:10 17. C-H and C-C of propane activation assisted by CO_2 for propylene and ethylene using bimetallic catalysts, *Dr. Emad N. Al-Shafei, Saudi Aramco R&DC*

11:30 18. CO_2 capture and conversion analysis technology development: TRL 1 -to- TRL 9
Dr. Ali S. Al-Hunaidy, Saudi Aramco R&DC

11:50 Prayer & Lunch Break

SESSION SIX   WASTE AND CO_2 UTILIZATION   ***Chairman: Prof. M. Mozahar Hossain, KFUPM, CHE***

13:00 19. Ceria-catalyzed synthesis of aliphatic polycarbonate diols from atmospheric flow carbon dioxide and diols without using dehydrating agents, *Prof. Keiichi Tomishige, Chairman of JPI Committee and Tohoku University, Japan*

13:20 20. Contaminants in plastic derived oil: A thorn in the flesh of plastic circularity?
Dr. Aaron Akah, Saudi Aramco R&DC

13:40 21. Conversion of rubber waste to liquid hydrocarbon fuel via thermal liquefaction process using solvent, *Dr. Nabeel Ahmad, IRC Refining & Advanced Chemicals, KFUPM*

14:00 *Closing Remarks, Symposium Ends*

Each presentation includes 5-minutes Q&A

JPI Direction along Carbon Neutrality and Introduction of Our Research Trend for Use of NanoTerasu, Next Generation Synchrotron Radiation Facility

Atsushi Muramatsu, Prof., Dr.,
Director of
International Center for Synchrotron Radiation Innovation Smart
Tohoku University
2-1-1, Katahira, Aoba-ku, Sendai 980-8579, Japan
Tel & Fax: +81-22-217-5163
E-mail: mura@tohoku.ac.jp

Abstract

Japan has set a target of carbon neutrality by 2050 so that all kinds of technologies and policies will be incorporated. We are now preparing an environment for research and technology development that can be expected for innovation. The Japan's 6th Strategic Energy Plan presented on October 2021 has claimed that the petroleum and petrochemical fields will continue to be important in the transportation field and material applications in spite of the requirement of completely reforming. We believe that the significance of using hydrocarbons will not change, but the introduction of carbon recycling technology is essential. Since catalysis science plays a major role in the production and utilization of hydrocarbons, we expect innovations in the future catalyst design through the complete evaluation of structure and properties of catalysts. I will now introduce our research project to understand the relationship between the structure and function of catalysts at the atomic and molecular level with use of next generation synchrotron radiation facility, NanoTerasu. Novel synthesis method for zeolite materials has been developed in our laboratory, which is based on mechanochemical reaction of precursor solid oxides, containing SiO_2 and the other metal (hydr)oxides, to give amorphous mixed oxides, and successive hydrothermal synthesis through conversion from them via solution phase. Synchrotron radiation X-ray spectroscopy gives us atomic-level evaluation of the zeolite frameworks.

1. Carbon Neutrality

1-1. 2050 Carbon Neutrality Declaration

In October 2020, Former Prime Minister Suga declared that Japan will aim to realize a decarbonized society by 2050, in other words to reduce greenhouse gas emissions to net zero. Japan's roadmap to tackle the challenge of climate change is bold and ambitious. It is marked by three key milestones. Firstly, Japan's commitment under the United Nations Climate Change Convention to reduce greenhouse gas (GHG) emissions by 26% from 2013 levels by 2030. The second milestone is to promote the development of innovative technologies by 2050 that enable Japan to contribute to the reduction in accumulated atmospheric CO2 globally to "Beyond Zero". The third and most ambitious milestone unveiled by Prime Minister Suga Yoshihide on October 26, 2020, calls for Japan to achieve net zero GHG emissions by 2050. This bold pledge sets Japan on a course to become Carbon Neutral in 30 years.

Japan's approach is based on three key principles

1. Promote Innovation & Technology as the agents of change in tackling the challenges of global warming.
2. Promote Green Finance to support the development of innovation and new technologies.
3. Support greater International Cooperation for business-led adoption of innovative green technologies.

1-2. Japan's energy policy

On the premise of safety, we will prioritize the stable supply of energy, realize a low-cost energy supply by improving economic efficiency, and at the same time strive for compatibility with the environment. Environmental measures shift from costs to growth opportunities, fossil resources shift from effective utilization to decarbonization (however, conversion requires time, cost, and innovation).

Japan's energy policy, etc.

Long-term energy supply and demand outlook	July 2015
The 5th Strategic Energy Plan	July 2018
Innovative Environmental Innovation Strategy	January 21, 2020
Former Prime Minister Suga Policy Speech "2050 Carbon Neutral"	October 26, 2020
Former Prime Minister Suga Climate Summit "2030 GHG reduction by 46% compared to 2013"	April 22, 2021
Green Growth Strategy Accompanying Carbon Neutrality in 2050	June 18, 2021
The 6th Strategic Energy Plan	October 22, 2021

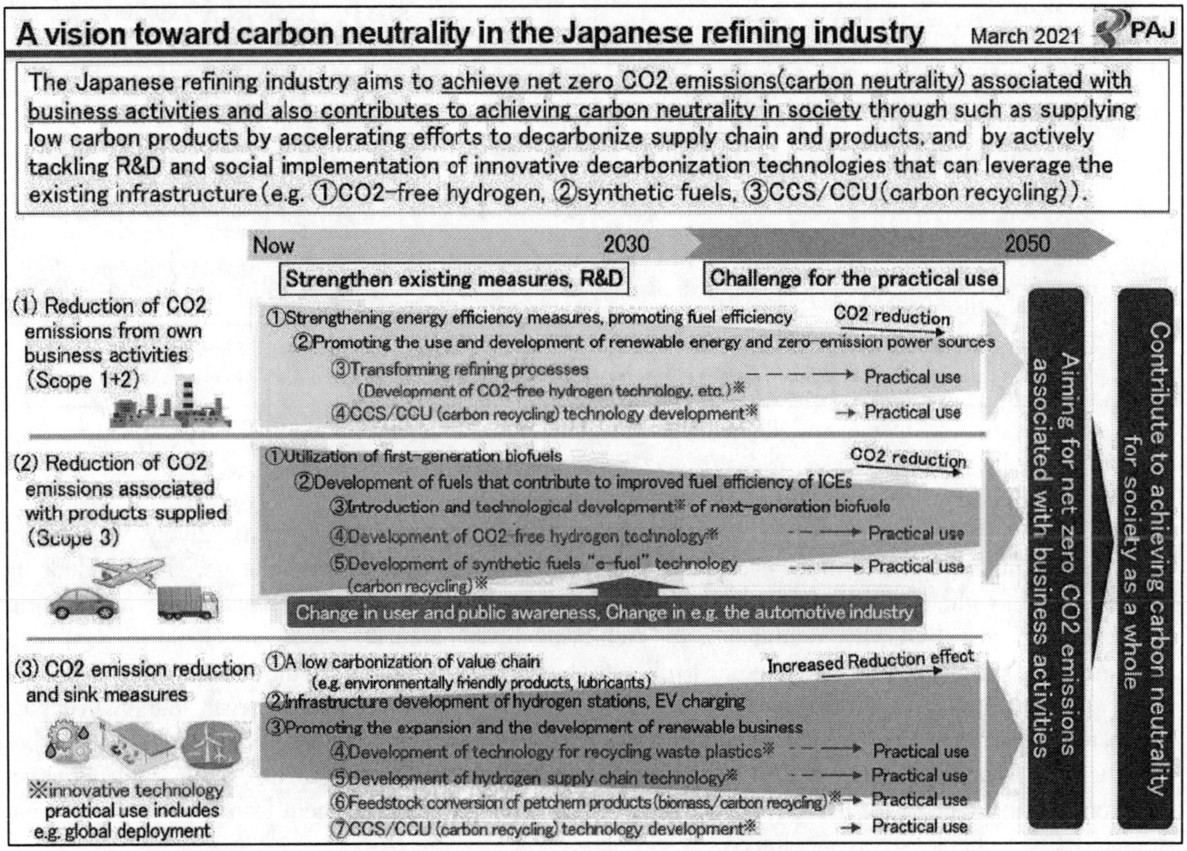

Figure 1. Vision of PAJ toward Carbon Neutrality for the Oil Industry

1-3. Petroleum Association of Japan

Based on these governmental actions, the Petroleum Association of Japan (PAJ) announced its Vision toward Carbon Neutrality for the Oil Industry (March 2021). It shows that the oil industry will work to achieve carbon neutrality based on the following three pillars:

1. reduction of emissions associated with its business activities,
2. reduction of emissions associated with the products it supplies,
3. measures to reduce CO2 emissions and carbon sinks.

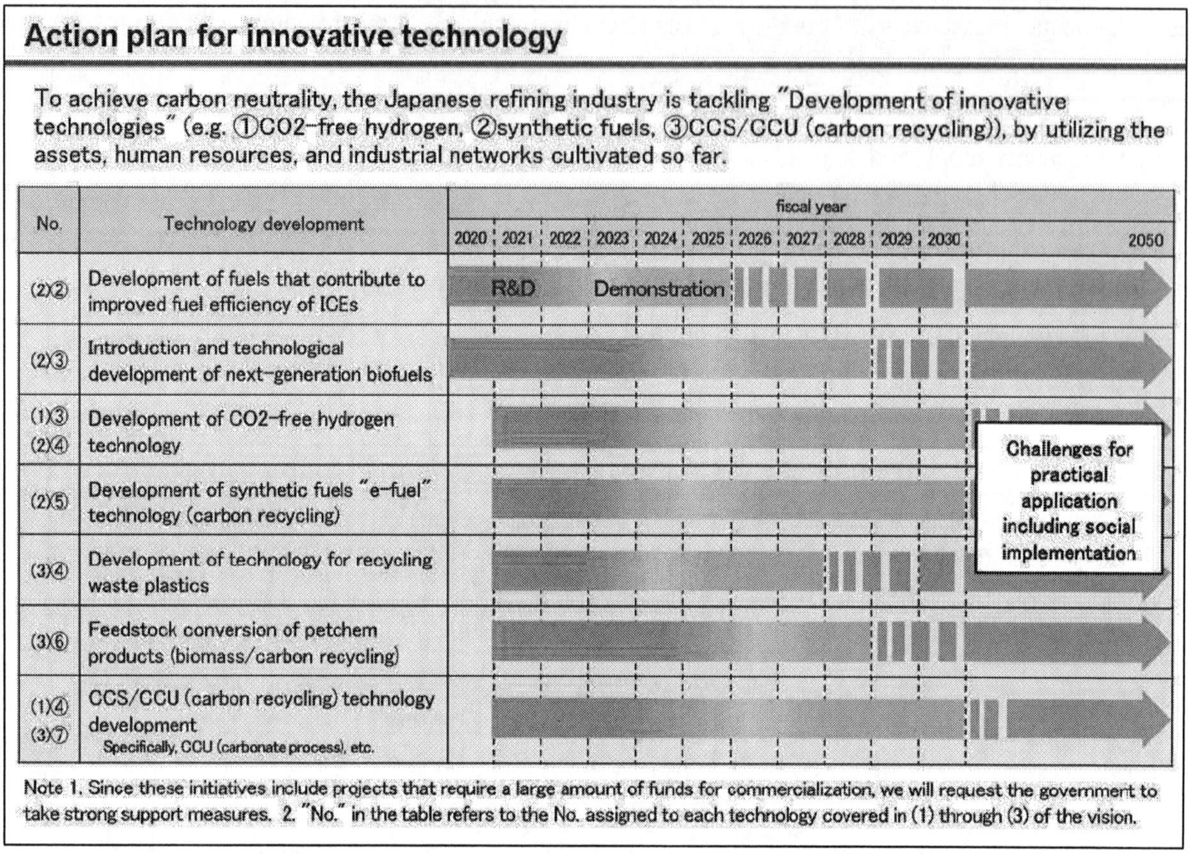

Figure 2. PAJ action plan for innovative technology

1-4. Japan Oil, Gas and Metals National Corporation

Japan Oil, Gas and Metals National Corporation (JOGMEC) has been actively working on developing technology and resource energy for decarbonization to tackle climate change. These include the development of carbon dioxide (CO2) sequestration/recovery, metal recycling, and geothermal technologies and resources. The "JOGMEC's Initiative for Achieving Sustainable Development Goals (SDGs)" was announced in December 2018. "Responding to climate change and contributing to the construction of a cyclical society" was one of the five action pillars emphasized, and the policy showed that the organization will engage in concrete measures against climate change. In July 2020, the "Technical Business Strategy for a Low-Carbon Society" was formulated, and a "CCS Group" was also established within JOGMEC, during which a policy for strengthening support for low-environmental-impact oil and gas development using carbon capture and storage (CCS) technology was announced.

In particular, JOGMEC will promote efforts for decarbonized fuels, such as hydrogen and ammonia. JPGMEC will strengthen support in fields where the technical knowledge of JOGMEC can be directly utilized, such as blue hydrogen and ammonia. We will also contribute to the construction of hydrogen and ammonia supply chains and market expansion, while utilizing our relationship with resource-rich countries.

JOGMEC will strengthen efforts for fossil fuel decarbonization technologies, such as Carbon dioxide Capture and Storage (CCS) and carbon recycling. JOGMEC will specifically maximize the use of technical knowledge of JOGMEC on the subsurface and surface characteristics and strengthen financial and technical support for Japanese companies in securing suitable sites for CCS.

1-5. Janan Petroleum Institute

Here, Japan Petroleum Institute (JPI) will explain the direction of carbon neutrality. As we have explained so far, Japanese policy and industry believe that petroleum, or liquid hydrocarbons, will be a necessary source of energy and raw materials even in the carbon-neutral era. Electric power and hydrogen generated by renewable energy are considered to be the main players in the carbon neutral era, but liquid hydrocarbons with high energy density and excellent portability are essential energy in fields such as transportation. However, hydrocarbons emit CO2 not only as fuel but also after being used for various purposes. CO2 will be emitted if the waste plastics after the use of plastics are disposed of by combustion. Therefore, there is a demand for the use of hydrocarbons after introducing carbon recycling technology.

However, there are many challenges. Typical examples include a significant increase in energy costs, transition period response, further technological improvement, and decentralization of renewable energy. First, there is a significant increase in energy costs. Compared to fossil resources such as oil, natural gas, and coal, hydrocarbons made from renewable energy are expensive raw materials. Therefore, it is necessary to utilize existing petroleum refining and petrochemical facilities in order to reduce costs as much as possible. Leveraging existing equipment also helps address the second transition period. And the most important thing for us is the further improvement of technology. We need to aim to improve basic research and core technologies that will lead to innovation. Finally, natural energy exists dispersedly. Unlike fossil resources, solar and wind power are ubiquitous, and there is very little energy in one place. For this reason, a strategy that matches the characteristics of natural energy is required, and the development of small-scale equipment that can handle less energy is also important.

The Japan Petroleum Institute aims to contribute to the resolution of issues in the carbon-neutral era by making the most of our unique characteristics. The Japan Petroleum Institute is a forum for interaction between industry and academia in the areas of technology and industry related to petroleum and petrochemicals. It is a place where companies and universities can share information and exchange opinions about their respective needs and seeds. It is the origin of open innovation for carbon neutrality.

The JPI's efforts for "carbon neutrality" are as follows.

First, we are promoting the establishment of basic technology for the production of CO2-free liquid hydrocarbons. We will promote the sharing of technical information developed and owned by our members. In addition, we will match the seeds of academia with the needs of industry toward 2050, and contribute to CO2-free technology in petroleum refining and petrochemical processes, and manufacturing technology of CO2-free products using petroleum refining and petrochemical facilities. to proceed. We will also work to build basic technologies that take advantage of regional characteristics and to achieve carbon neutrality within the region.

Finally, we will provide a correct evaluation from the scientific aspect of the business aimed at carbon neutrality. Technological development is progressing in the world by taking all possible measures to achieve carbon neutrality.

Finally, I will explain an example of our efforts.

The Japan Petroleum Institute has members with expertise in petroleum development/exploration, petroleum refining, petrochemicals, petroleum product manufacturing/utilization technology, and transportation technology. In addition, there are researchers who possess basic technologies such as chemical engineering, catalytic chemistry, interfacial (surface) chemistry, and electrochemistry. We believe that the application of new carbon-neutral hydrocarbon production technology and existing carbon-neutral technology to petroleum refining and petrochemical equipment will progress by organizing these human resources and existing and cutting-edge technologies and creating a database.

In order to promote carbon neutrality, basic technology must be established. Next, I would like to introduce my research as well as the basic technologies necessary to achieve carbon neutrality.

2. Novel synthesis method for zeolites and their evaluation with use of synchrotron radiation

2-1. Mechanochemical route

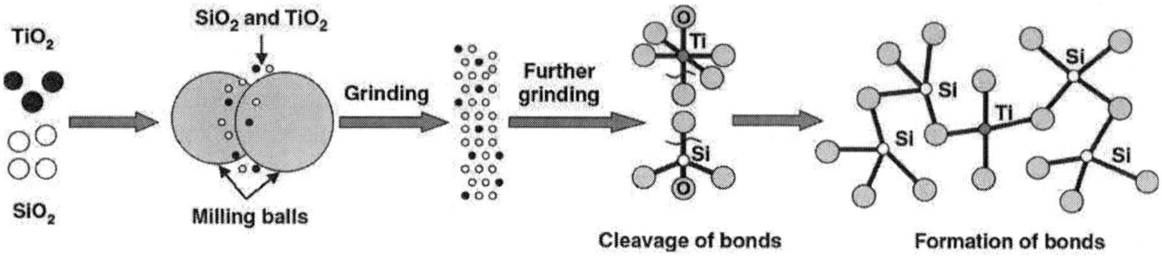

Figure 3. Schematic illustration of the precursor preparation via the mechanochemical process for Titano-silicate (TS-1)

We have already reported the development of a new synthesis method for TS-1 utilizing mechanochemical reaction [1-3]. This mechanochemical synthesis method employs titania and silica powders as starting materials, which are allowed to react mechanochemically to be a silica–titania composite as a precursor to TS-1. Figure 3 illustrates the precursor preparation process using the mechanochemical reaction occurring at the grinding using a planetary ball mill. By the rotation and the revolution of a milling pot, balls installed inside the pot hit each other to produce powerful impact and/or friction energy according to operational conditions, such as the milling time and the disk rotation speed etc. [4]. This mechanical energy promotes the degradation of crystalline structure, the cleavage of bonds, and the formation of new bonds between silicon and titanium species, resulting in the formation of a uniform silica–titania mixture. Thus prepared silica–titania mixture is hydrothermally treated in the presence of organic structure-directing agent (SDA) to be crystallized into TS-1 materials.

Concerning the plausible mechanism of this two-step mechanochemically assisted procedure, it can be said that the chemical potentials of the complexes containing Si and the substituting metal atoms in a liquid phase are kept constant throughout the hydrothermal process owing to the solid–liquid equilibrium, which could be aiding the formation of more homogeneous zeolite frameworks compared to zeolites synthesized via the conventional one-pot hydrothermal process using dissolved starting reagent(s).

2-2. Preparation of MFI Zeolites Substituted Isomorphously by Fe [5, 6]

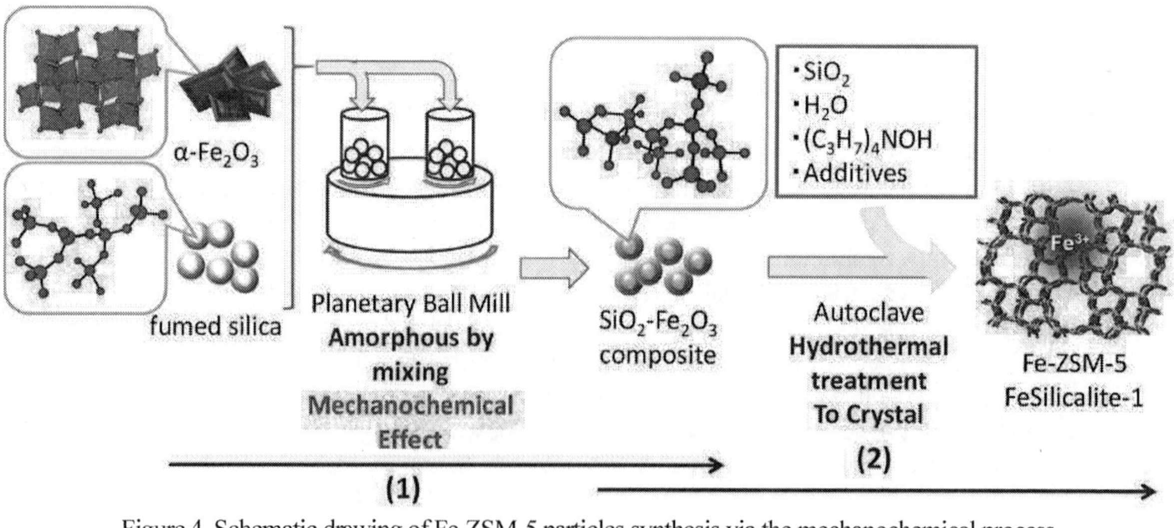

Figure 4. Schematic drawing of Fe-ZSM-5 particles synthesis via the mechanochemical process

The present mechanochemical approach can potentially yield an MFI-type zeolite substituted isomorphously by Fe with better homogeneity. Here, we aimed to prepare [Fe]-MFI zeolites via the mechanochemically assisted procedure.

Initially, 60 mmol fumed silica (Aerosil 200V, Evonik Industries) was pulverized using a planetary ball mill (Pulverisette 7, Fritsch) with 1.5 mmol α-Fe_2O_3 or 3.0 mmol in a Si_3N_4 pot in the presence of seven Si_3N_4 balls (diameter 15 mm). Pulverization was performed at 600 rpm, with an interval of 15 min after each milling run for 15 min. The total milling time for the preparation of a composite containing Si and Fe was varied from 0 to 48 h. The subsequent hydrothermal synthesis to convert the prepared composites into MFI zeolites was conducted using a Teflon-lined high-pressure reactor (Parr Instrument Company, inner volume 23 mL). The ball-milled SiO_2–α-Fe_2O_3 composite, silica gel (Carplex BS-304F, DSL. Japan), an aqueous solution of tetrapropylammonium hydroxide (TPAOH, 40 wt %), an aqueous HCl solution, an aqueous NaCl solution, and deionized water were charged into the reactor. The molar ratio of Si/FeTPAOH/HCl/NaCl/H2O was adjusted to 1.0/0–0.02/0.5/0–0.3/0–0.5/10–45. The mixture was stirred at room temperature for 0.25–168 h, preheated at 50–100 °C for 0–72 h at a tumbling speed of 10 rpm, and then treated hydrothermally at 160 °C for 3–168 h at the same tumbling speed. The solid phase was washed with deionized water using a centrifuge, dried, and calcined in air at 540 °C for 12 h. To generate Brønsted acid sites over each sample surface, the resulting powder was treated four times in an aqueous NH_4Cl solution (1 M) at 60 °C for 1 h. After washing 10 times with deionized water using the centrifuge, drying at 60C for 24 h, and calcination in air at 540 °C for 12 h, an MFI-type zeolite with H+ counter cations was obtained.

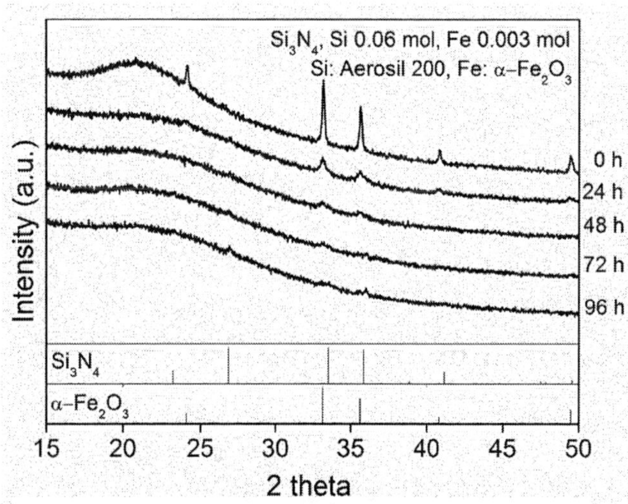

Figure 5. Effects of milling time on the structure of oxide mixtures determined by XRD measurement

In the XRD patterns of the mixtures of SiO_2 and α-Fe_2O_3 (Figure 5), the distinct diffraction peaks at 2θ values of 24.1, 33.2, 35.6, 40.9, 49.5, and 54.1°, all of which are attributed to α-Fe_2O_3, observed before the milling process were diminished upon an increase in milling time to 48 h. This gradual decrease indicated that amorphization of α-Fe_2O_3 occurred during milling. For milling times of 36 and 48 h, the tiny diffractions detected at 2θ values of 27.0 and 36.0° were assignable to β-Si_3N_4. The appearance of these peaks pointed to the slight β-Si_3N_4 contamination in the milled samples from the pot and balls used for pulverization, which was consistent with the XRF measurements (vide supra). In the UV–vis spectra in Figure 6, regardless of milling time, all of the milled samples exhibited the prominent absorption band centered at 254 nm that derives from isolated and tetrahedrally coordinated Fe^{3+} species [7, 8] incorporated into the SiO_2 network as a result of the mechanochemical reaction between SiO_2 and α-

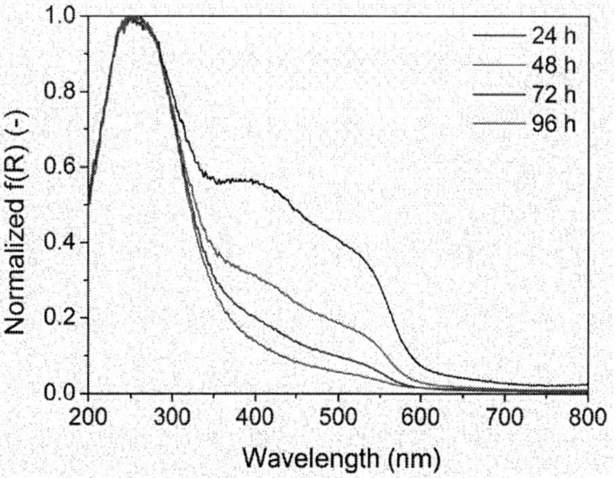

Figure 6. Effects of milling time on the nature of Fe species, examined by DR-UV–vis spectroscopy. The spectra were normalized at the maximum absorption.

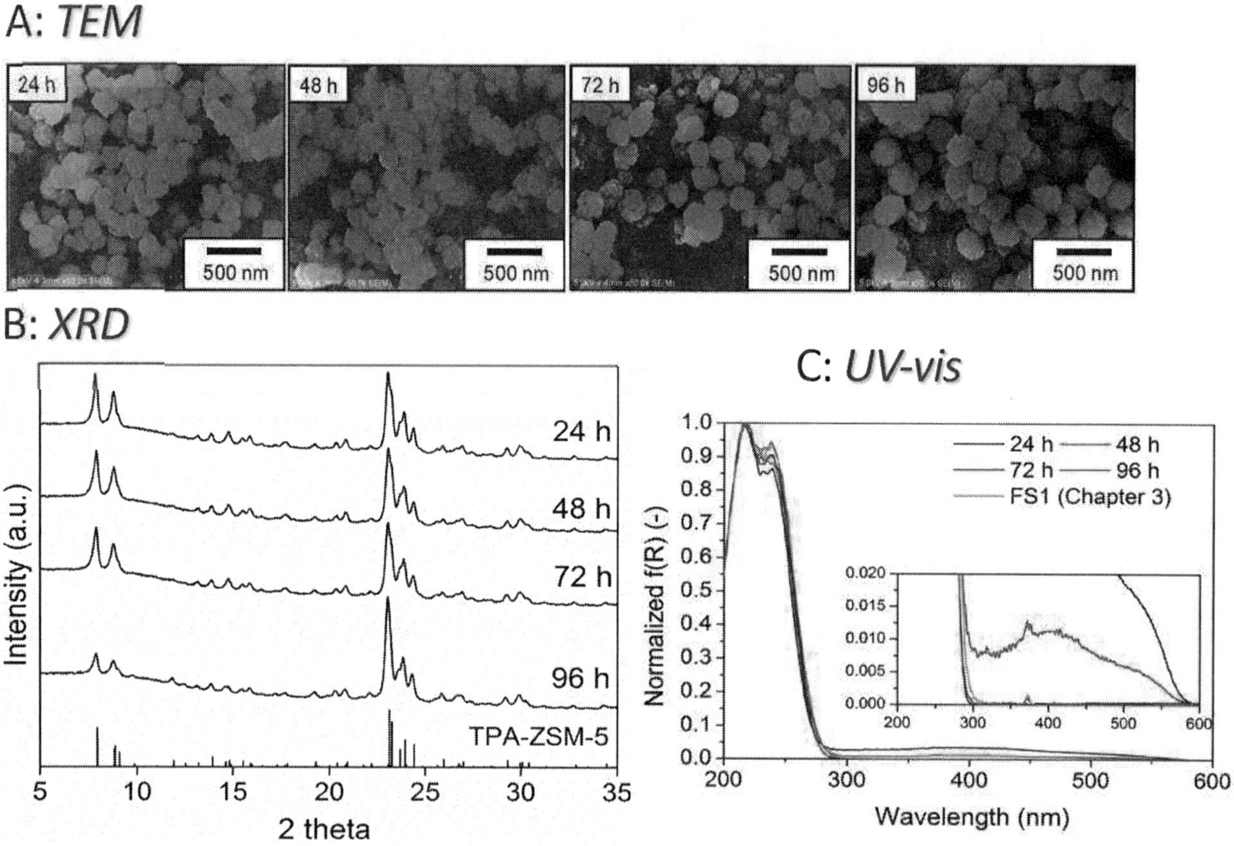

Figure 7. Effects of milling time during SiO_2–α-Fe_2O_3 preparation on textural properties of as-synthesized MFI zeolites formed hydrothermally from the composites: (A) SEM images; (B) XRD patterns; and (C) DR-UV–vis spectra. In B, the reference of ZSM-5 containing TPA cations (TPA-ZSM- 5) is from the database of zeolite structures, adapted with permission from Baerlocher, C, et al. [10] The inset of C represents the expanded spectra. Hydrothermal conditions: Si/Fe = 50 (SiO_2–α- Fe_2O_3 composites milled for 12–48 h were used); TPAOH/Si = 0.5; HCl/Si = 0.2; Na/Si = 0.1; H_2O/Si = 30; aging at room temperature for 48 h; and hydrothermal treatment at 160 °C for 120 h.

Fe_2O_3. Since the other absorption bands at 400–600 nm due to α-Fe_2O_3 weakened with increasing milling time, the relative intensity of the peak at 254 nm increased. This trend suggests that longer planetary ball-milling times lead to the collapse of the α-Fe_2O_3 crystalline structure as well as the incorporation of Fe^{3+} species into the SiO_2 network. The mechanochemical reaction between two oxides yielded an amorphous composite, while a similar technique, so-called mechanical alloying, was reported to yield crystalline alloys. [9] The contrasting results of these two mechanical-force-induced reaction systems could be due to the presence/absence of oxygen atoms in the starting reagents and desired products since oxygen leads to a more complex chemical bonding network and prevents the formation of an ordered structure.

The four (24, 48, 72, and 96 h) mechanochemically prepared amorphous SiO2–α-Fe_2O_3 composites were employed as reagents for the second step of the mechanochemical method, i.e., the hydrothermal treatment, to synthesize [Fe]-MFI zeolites (see Figure 7). Prior to calcination for removing the TPA cations from the samples, the as-synthesized materials were characterized by SEM, XRD, and UV–vis measurements. Spherical nanoparticles were observed in the SEM images (Figure 7A). The samples exhibited diffraction patterns that were identical to the reference ZSM-5 containing TPA cations [10] (Figure 7B), which confirmed the formation of an MFI-type zeolite framework through the hydrothermal treatment using each amorphous composite. It should be noted that the baseline elevation found in the 2θ range of 5–15° is probably caused by diffuse X-ray scattering since such phenomena were also observed in our previous reports investigating a variety of functional materials. [11-13] In the case of [Fe]-MFI samples prepared from amorphous composites at milling times of 36 and 48 h, only the characteristic peak at 210− 250 nm due to the tetrahedrally coordinated Fe^{3+} species in the zeolite framework was observed in the DR-UV−vis spectra (Figure 7C). [14-21] The Si/Fe ratios of the solid samples remained almost unchanged during hydrothermal treatment. Thus, the Fe^{3+} species originally present in the SiO_2 network were successfully taken into the resulting zeolite framework; in other words, small fragments of ferrisilicates were presumably dissolved from the amorphous composite, behaving as building units toward the construction of a heteroatom-substituted zeolite framework during the hydrothermal process. Along with the peak at 210− 250 nm, additional absorption bands at 300−600 nm, which are assignable to isolated octahedrally coordinated Fe and/or Fe_xO_y clusters outside the zeolite framework (250−350 nm), larger Fe clusters (350−450 nm), and bulk Fe species (>450 nm), [14-21] appeared for the [Fe]-MFI synthesized from composites that were ball-milled for 12 and 24 h (Figure 7C). In these cases, a part of the Fe species remaining as α-Fe_2O_3 outside the SiO_2 network owing to insufficient ball-milling (see Figure 6) formed clusters and nanoparticles on the zeolite surfaces, along with isolated tetrahedrally coordinated Fe species in the zeolite framework. The formation of the extraframework Fe species also indicated that the incorporation of Fe species into the SiO_2 network via the mechanochemical process was key to the preparation of [Fe]-MFI. Thus, the amorphous composite ball- milled for 48 h was employed for the hydrothermal process in the following experiments because, in this case, the resulting zeolites did not contain any extraframework Fe species.

But, the discussion on the coordination of Fe into the zeolite framework has been carried out basically under the speculation based on the results from many analysis methods, because we have never been obtained the direct observation technique to the zeolite framework itself.

2-3. Comprehensive understanding of well-defined zeolite synthesis and detailed structural organization

The answer to that undissolved problem is this project. This research project is funded by the Ministry of Education, Culture, Sports, Science and Technology of Japan, Grants-in-Aid for Scientific Research S. Here, we are looking at the use of the world's top-level zeolite synthesis technology, which we have already developed, and XAS etc. of the next-generation synchrotron radiation NanoTerasu. By synthesizing zeolites with precisely controlled atomic arrangements and using them for structural analysis, we establish novel structural analysis methods and synthesis methods for structural arrangement-controlled zeolites.

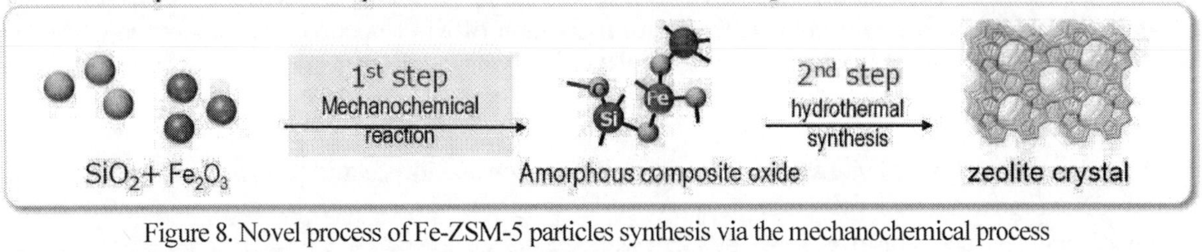

Figure 8. Novel process of Fe-ZSM-5 particles synthesis via the mechanochemical process

Feasibility study is underway at SPring-8, because NanoTerasu is now under contraction. We have studied the effect of ball-milling time, that is, the period of mechanochemical reaction in 1st step of Fig. 8, on the coordination number of Fe by using XAFS of SPring-8. From our experience of the observation of as-ball-milled samples, with increasing milling time, the sample color tended to fade. We imagine that Fe-O-Si bond can be formed during the mechano-chemical reaction, and then Fe_2O_3 crystalline phase turns to amorphous structure because as a result of substitution of heteroatoms to Si-O-Si bonds, the crystal is highly strained and becomes amorphous. Focusing on Fe atoms, in the Fe_2O_3 crystal, Fe is octahedral, but when Si-O-Si is substituted with Fe, it becomes tetracoordinated. We have tried to evaluate Fe coordination during the mechanochemical reaction.

Figure 9 shows the tracking of the synthesis process of Fe-containing amorphous complex oxides by mechanochemical reactions using XAFS. The figure on the left of Fig. 9 demonstrates the X-ray Absorption Near Edge Structure (XANES) spectrum. The small part at around 7107 eV is called the pre-edge peak, and

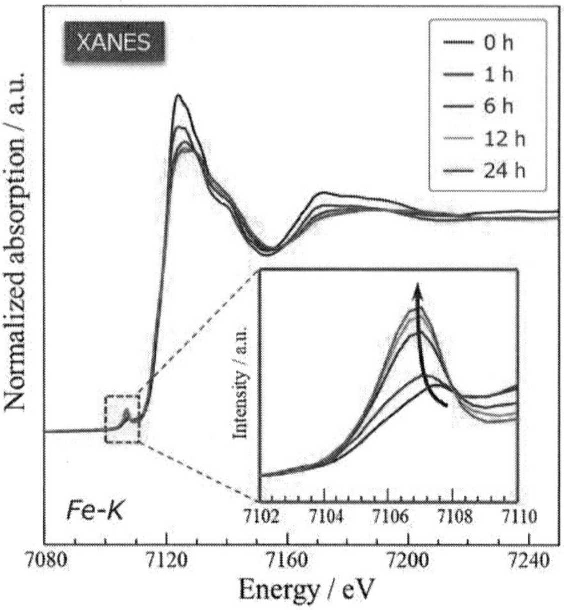

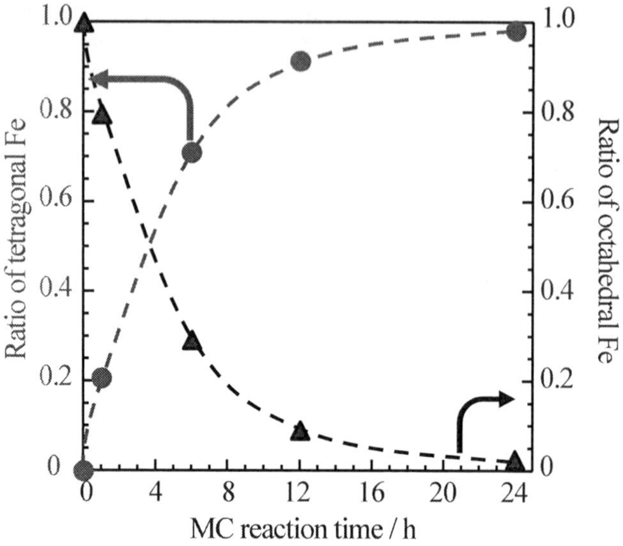

Figure 9. Left: XAFS spectra of K-edge XANS spectrum of Fe and Right: Ratio of tetragonal Fe and octahedral Fe calculated from pre-edge peaks in XANES region.

you can see the absorption depending on the coordination number of Fe. The right figure in Fig. 9: Ratio of tetragonal Fe and octahedral Fe calculated from pre-edge peaks in XANES region. As shown in right-side of Fig. 9, tetragonal Fe increased at the expense of octahedral Fe with mechanochemical reaction time. We also obtained the evidence that this Fe-O-Si structure is directly transferred to the produced zeolite. Namely, Si-Fe bimetallic oxide or hydroxide complexes were once dissolved in aqueous solution phase, and then Fe-MFI zeolite crystal was formed via reprecipitation of the complexes as a precursor. Detailed analysis is currently in progress.

In this project, we have already succeed to quantitatively evaluate the proportion of tetracoordinated Fe species from the analysis of the pre-edge peak. Also, shapes of XANES spectrum were found to differ depending on the framework structure. One can consider that they reflect the difference in the electronic state due to the bond angle. We will investigate direct observation of M-O bonds by absorption and emission spectroscopy at different absorption edges (orbitals) and detailed state selection by resonance emission spectroscopy in the next step.

3. Introduction on Next Generation Synchrotron Radiation Facility, NanoTerasu

Next generation synchrotron radiation facility, NanoTerasu, is now under construction on our Aobayama campus, as shown in Fig. 10. Compared with SPring-8, NanoTerasu will supply remarkably high energy soft X-ray with ca 100 times brighter radiation.

Figure 10. Photo of NanoTerasu on August, 2022

Figure 11 shows NanoTerasu Facility Specification. Since its circumference is only 350 m, which is around 1/4 of that of SPring-8, you can see that it is very compact but the performance is much more excellent.

As shown in Fig. 12, for

Beam energy	3 GeV
Current	400 mA
Lattice	4-bend achromat
Natural emittance	1.1 nm.rad
Circumference	348.8 m
Cell	16
Long straight / ID length	5.40 m / 4.2 m
Short straight / ID length	1.66 m / 0.7 m
Beamline (long straight)	14
Beamline (short straight)	12

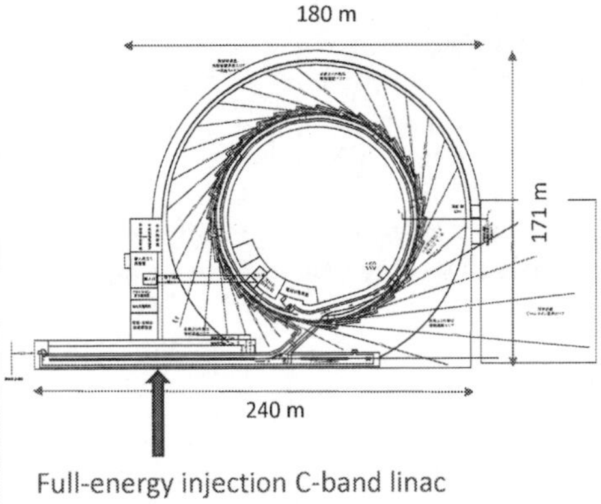

Figure 11. NanoTerasu Facility Specification

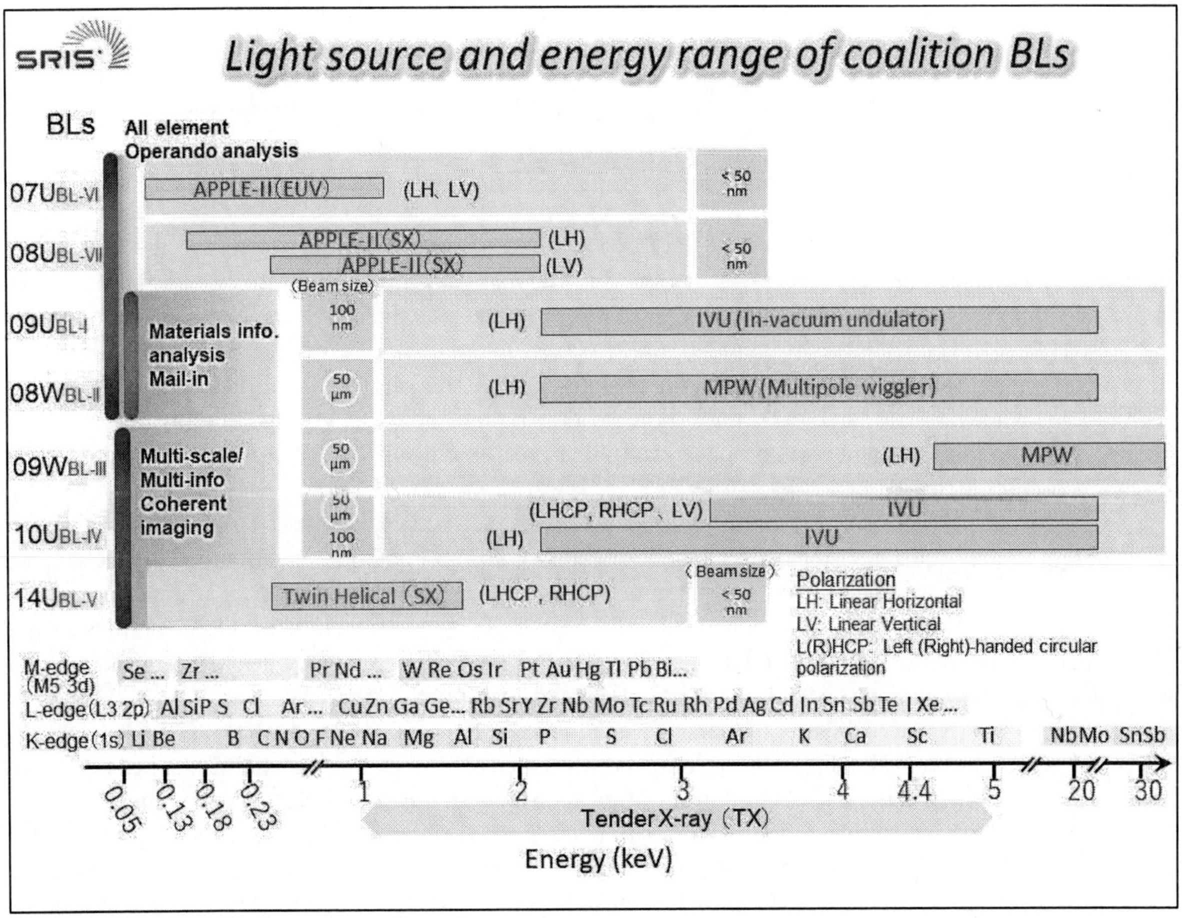

Figure 12. Light source and energy range of Coalition Beamlines of NanoTerasu

NanoTerasu, it is possible to observe elements with small atomic numbers, such as Li, B, etc. For example, Li cation movement in battery can be visualized. Change in valence and/or oxidation number of B can be monitored in situ during reactions. Since the concept of NanoTerasu is based on industrial use, how to use Coalition Beamlines (BLs) of NanoTerasu is quite unique in the registration of Coalition Membership. There is no need to write a troublesome application form of Grant-in-Aid for Scientific Research, you just make a reservation in the previous month before use.

Figure 13 shows the layout of the first 10 beamlines to be built in 2023, including 3 public Beamlines and 7 Coalition Beamlines, for Coalition numbers, mainly in the industrial use. In this regard, the Coalition

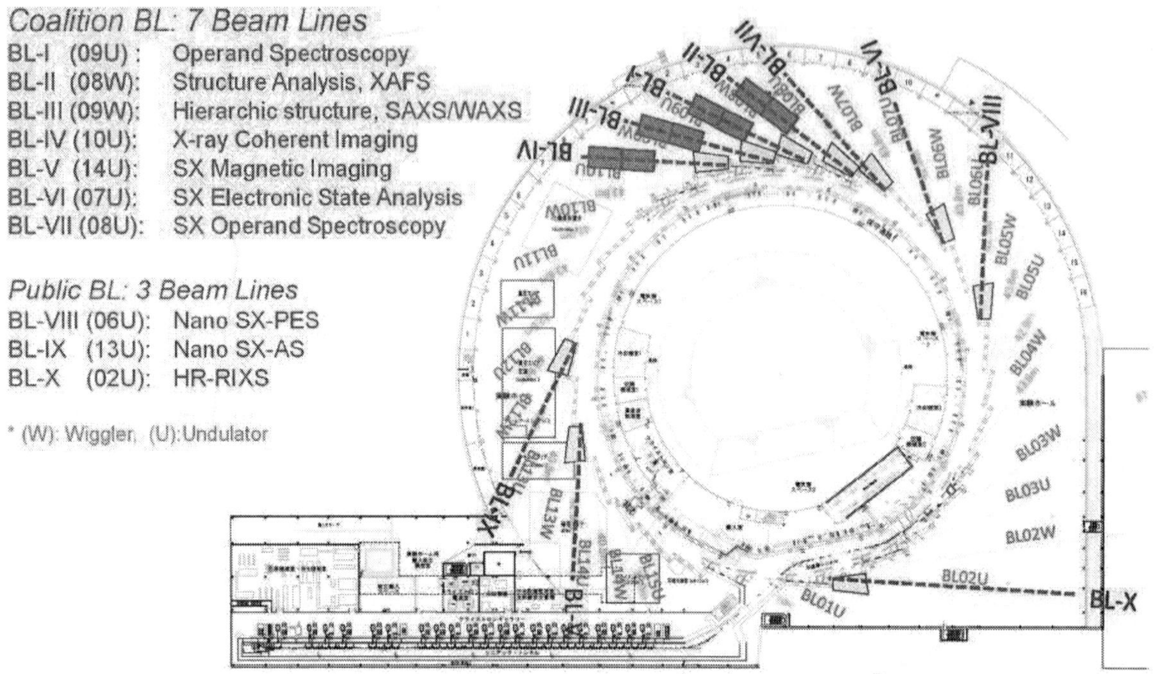

Figure 12. Layout of the first 10 beamlines to be built in 2023

membership is mainly purchased by private companies. The main incentive for this membership is the right to use any Coalition beamline for 200 hours per year starting 10 years after the year NanoTerasu started operation.

Hence, Tohoku University founded a new center, International Center for Synchrotron Radiation Innovation Smart (SRIS), for the purpose of the promotion of the fusion of different fields, such as industry, with diverse academic fields and leads innovation creation. As a result, when the operation will start in 2024 spring, it will be used smoothly and produce useful results so that it will contribute to the formation of the research complex in Sendai.

The roles of SRIS are to lead full-scale cooperation in science and engineering of end stations in BLs from design to their utilization for industry, in corporation with the Photon Science Innovation Center. Our center will also serve as a hub for collaboration between overseas research complexes. For an international scale, we will perform research, education, and human resources development utilizing these facilities through the cooperation both of universities and academic organizations. We have already held the workshop for the utilization of next-generation synchrotron radiation facilities for 3 times from 2019.

Four missions of SRIS, are, 1. Leading academic research / industry-academia collaboration utilizing next-generation synchrotron radiation; Pioneering new research areas that also incorporate the perspectives of social science and economics, 2. Building an innovation system through industry-government-academia collaboration; Creating new knowledge and fusion of fields that contribute to industry and society, 3. Formation of international university synchrotron radiation alliance; Strengthening international competitiveness by forming research networks with overseas synchrotron radiation facilities and collaborative research institutions, and 4. Human resources development utilizing synchrotron radiation facilities; Leading young researchers who will lead the development of new industries by creating a new field with a broad perspective.

We are planning to use Coalition Beamlines in NanoTerasu for the exact evaluation of frameworks of zeolites we synthesize by novel methods. Hence, our collaborators are now investigating novel evaluation procedures and processes to be established by the operation start of NanoTerasu.

4. Conclusions

Today, I summarized JPI direction along Carbon Neutrality in Japan. In addition to this, I described the introduction of our research project, and next generation synchrotron radiation facility, NanoTerasu. Although the three topics seem completely separate, they are actually closely related.

5. References

1. K. Yamamoto, S.E. Borjas Garcia, F. Saito, A. Muramatsu, *Chem. Lett.* **35**, 570 (2006).
2. S.E. Borjas Garcia, K. Yamamoto, F. Saito, A. Muramatsu, *J Jpn Petrol Inst* **50**, 53 (2007).
3. K. Yamamoto, S.E. Borjas Garcia, Muramatsu, *Micropor Mesopor Mat.* **101**, 90 (2007).
4. S.E. Borjas Garcia, K. Yamamoto, A. Muramatsu, *J Mater Sci* **43**, 2367(2008).
5. H. Kobayashi, M. Nakaya, K. Kanie, A. Muramatsu, *J. Environ. Sci. Eng. B*, **4**, 1 (2015).
6. M. Yabushita, H. Kobayashi, R. Osuga, M. Nakaya, M. Matsubara, S. Maki, K. Kanie, A. Muramatsu, *Ind. Eng. Chem. Res.*, **60**, 2079 (2021).
7. P. B. Amama, S. Lim, D. Ciuparu, Y. Yang, L. Pfefferle, J. Haller, *Phys. Chem. B*, **109**, 2645 (2005).
8. Y. Li, Z. Feng, Y. Lian, K. Sun, L. Zhang, G. Jia, Q. Yang, *Microporous Mesoporous Mater.*, **84**, 41 (2005).
9. C. Suryanarayana, *Mechanical alloying and milling. Prog. Mater. Sci.*, **46**, 1 (2001).
10. C. Baerlocher, L. B. McCusker, *Database of Zeolite Structures 2021* http://www.iza-structure.org/databases/. (accessed Jan 1, 2021).
11. M. Yabushita, H. Kobayashi, A. Neya, M. Nakaya, S. Maki, M. Matsubara, K. Kanie, A. Muramatsu, *CrystEngComm*, **22**, 7556 (2020).
12. M. Yabushita, M. Yoshida, F. Muto, M. Horie, Y. Kunitake, T. Nishitoba, S. Maki, K. Kanie, T. Yokoi, A. Muramatsu, *Mol. Catal.*, **478**, No. 110579 (2019).
13. R. Suzuki, Y. Nishi, M. Matsubara, A. Muramatsu, K. Kanie, *ACS Appl. Nano Mater.* **3**, 4870 (2020).
14. S. Bordiga, R. Buzzoni, F. Geobaldo, C. Lamberti, E. Giamello, A. Zecchina, G. Leofanti, G. Petrini, G. Tozzola, G. Vlaic, *J. Catal.*, **158**, 486 (1996).
15. J. Peŕez-Ramírez, *J. Catal.*, **227**, 512 (2004).
16. M.S. Kumar, M. Schwidder, W. Grünert, A. Brückner, A. *J. Catal.*, **227**, 384 (2004).
17. J. Peŕez-Ramírez, J.C. Groen, A. Brückner, M.S. Kumar, U. Bentrup, M.N. Debbagh, L.A. Villaescusa, *J. Catal.*, **232**, 318 (2005).
18. A. Zecchina, M. Rivallan, G. Berlier, C. Lamberti, G. Ricchiardi, *Phys. Chem. Chem. Phys.*, **9**, 3483 (2007).
19. C. Hammond, N. Dimitratos, J.A. Lopez-Sanchez, R.L. Jenkins, G. Whiting, S.A. Kondrat, M.H. ab Rahim, M. M. Forde, A. Thetford, H. Hagen, E.E. Stangland, J.M. Moulijn, S.H. Taylor, D.J. Willock, G.J. Hutchings, *ACS Catal.*, **3**, 1835 (2013).
20. K. Miyake, Y. Hirota, K. Ono, Y. Uchida, M. Miyamoto, N. Nishiyama, *New J. Chem.*, 41, 2235 (2017).
21. P. Xiao, Y. Wang, J.N. Kondo, T. Yokoi, *Appl. Catal., A*, **579**, 159 (2019).

Residue upgrading by hydroprocessing deasphalted oil

Koichi Matsushita

Central Technical Research Laboratory, ENEOS Corporation
Chidori-cho 8, Naka-ku, Yokohama 231-0815, Japan
E mail: matsushita.koichi@eneos.com

Abstract

The direction of heavy oil cracking in most Japanese refineries is in the utilization of the residue fluid catalytic cracking (RFCC) process. The sulfur content of low sulfur fuel oil (LSFO) should be reduced to below 0.5 wt% to comply with International Maritime Organization regulations. To achieve this, the residue hydrodesulfurization (RDS) unit must be operated under more severe conditions. For this study, a new RDS catalyst system to treat deasphalted oil (DAO) from the solvent de-asphalting (SDA) unit was developed. This system allows us to produce both RFCC feedstocks and LSFO from the vacuum residue (VR) fraction simultaneously and operate the refinery flexibly. DAO has a high molecular weight due to its coming from the VR fraction. However, the SDA process selectively reduces large molecules of vanadium and nickel, both catalyst poisoning materials. Such metal compounds in DAO are easy to remove, but due to their reactivity, the catalysts deactivate faster than conventional atmospheric residue (AR). To combat this, we developed a new RDS catalyst system for DAO treatment. This new system can suppress metal deactivation in the RDS unit and reduce the amount of nitrogen in the products, which in turn leads to an increase in RFCC conversion.

1. Introduction

Although demand for heavy oil products in Japan is seeming to decrease, the effective use of low-priced heavy crude oil will become increasingly necessary. Asphaltenes included in the VR of heavy oil are composed of sulfur, metals (vanadium and nickel), polycyclic aromatics, and long-chain alkyl groups. It is strongly agglomerated by hydrogen bonds, aromatic π - π bonds, acid-base bonds, etc. Therefore, it is very difficult to remove the sulfur and metals incorporated in the asphaltene molecules, and the catalytic active sites are poisoned and deactivated. To address this issue, an asphaltene relaxation method was proposed, but the direct removal of asphaltenes is more essentially an effective and efficient technique.

Therefore, this study focused on the SDA process for obtaining DAO by extracting and removing the asphaltenes in VR and developed an RDS catalyst system. By processing DAO in an RDS unit and a RFCC unit, more than 70% of the VR fraction can be converted into gasoline and chemicals resources such as propylene and BTXs (benzene, toluene, xylene), etc. (**Fig. 1**).

The supply of LSFO is required for bunker fuel oil in response to the sulfur regulation that the International Maritime Organization (IMO) started in January 2020. Since RDS product oil contains little asphaltenes, which tend to generate sludge, this technique is also extremely effective from the viewpoint of co-production of high-quality bunker fuel oil.

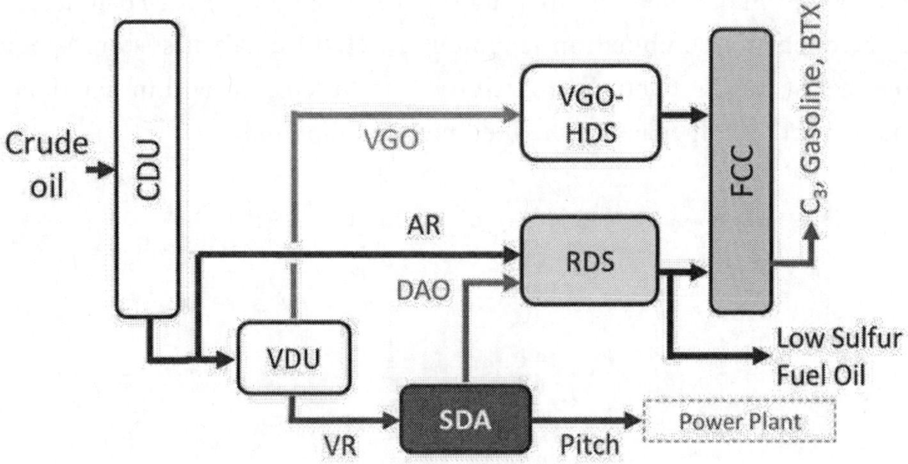

Figure 1. Bottom process flow of developed system

2. Detailed metal compound analysis in DAO and the reactivity

When an aging test of DAO as a feedstock was conducted with a conventional AR desulfurization catalyst system, it was found that the catalyst life was shorter than expected. To elucidate causes of the deactivation, gel permeation chromatography-inductive coupled plasma analysis was conducted. **Figure 2** shows the molecular weight distribution of vanadium compounds contained in AR and DAO.

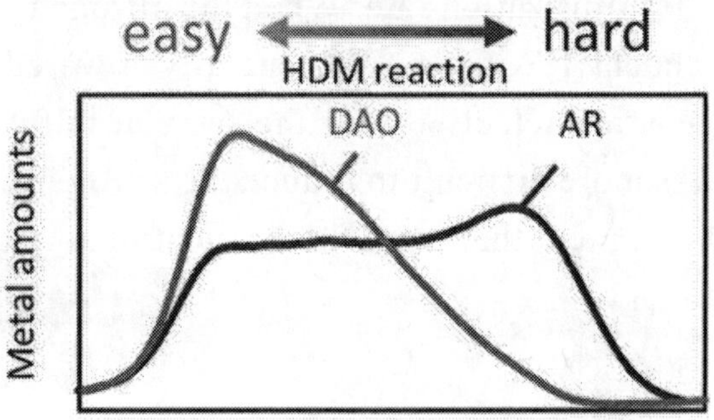

Figure 2. Vanadium compound distributions of AR and DAO

Since DAO is extracted from the VR fraction, it has a higher boiling point than AR. There is no difference in the reactivity of DAO and AR in hydrodesulfurizatoin (HDS, **Fig. 3a**) but a large difference in hydrodemetallization (HDM, **Fig. 3b**). In other words, it is confirmed that the HDM reaction proceeds more easily with DAO than with AR. The reason why the HDM reactivity of DAO is higher than that of AR, but the catalyst deactivates faster, is considered as follows. Because of the high HDM reactivity of DAO, metal deposition occurs near the inlet of the catalyst pores; thus, a deactivation mechanism that promotes pore mouth plugging has been proposed. To address this issue, we have established a combination technique of HDM catalyst system specialized for DAO treatment, which controls reactivity and diffusion within catalyst pores in accordance with the properties of the feed metal compounds.

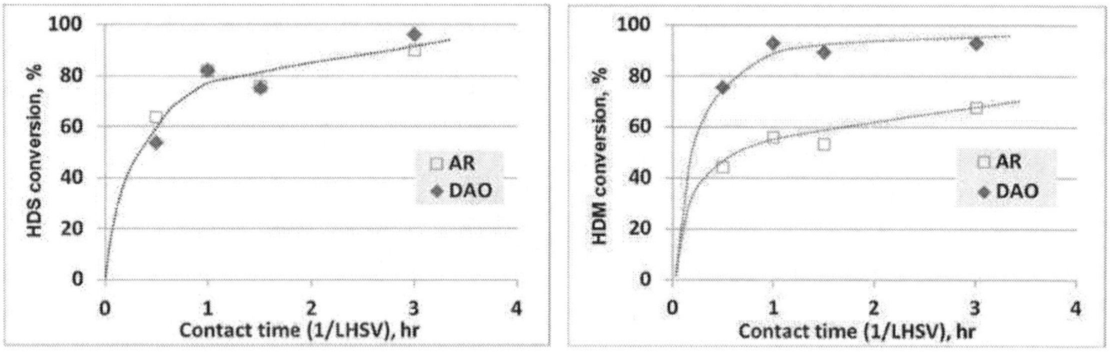

Figure 3. Reactivity differences of HDS (3a, left) and HDM (3b, right)

3. Development of highly selective hydrodenitrogenation (HDN) catalyst for RFCC reactivity enhancement

The RDS units in refineries are normally operated with a constant sulfur concentration in the product oil, and if the amount of nitrogen concentration in the RDS product oil needs to be reduced, it is necessary to improve the denitrogenation selectivity at the same desulfurization rate. When the nitrogen concentration is high, the reactivity of the subsequent RFCC unit is generally lowered. In our pilot plant, we compared the denitrogenation selectivity for the same desulfurization rate of DAO and AR. We found that it was more difficult to reduce the nitrogen concentration in the RDS product oil with DAO treatment than with the conventional AR treatment (**Fig. 4**).

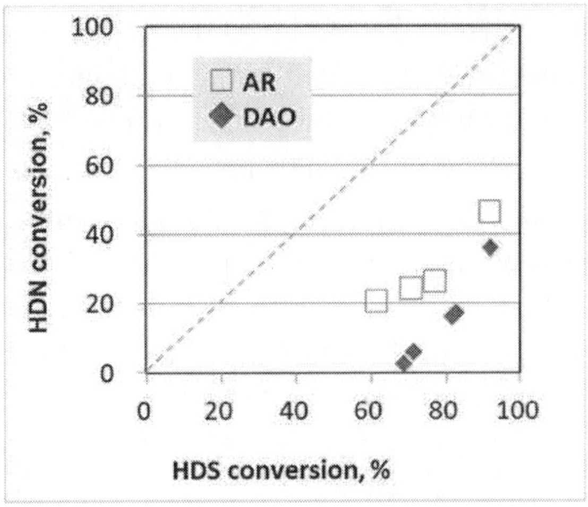

Figure 4. Relation between HDS and HDN reactivity

4. Design of catalyst system using catalyst life prediction simulator specialized for DAO treatment

In addition to the significant difference in HDM reactivity between DAO and AR, it was found that the exothermic state in the reactor was also different. It became possible to predict heat generation due to aromatic nuclear hydrogenation reaction from the structure of hydrocarbons through detailed structural analysis of DAO. Thus, we were able to accurately estimate the actual reaction temperature and establish a simulator that can accurately predict the reactivity and catalyst life against DAO processing[1)].

Using the simulator arranged for DAO treatment, the catalyst activity as well as the metal deposition on the spent catalysts, which is a factor in catalyst deactivation, could be accurately predicted. As a result, each catalyst can be loaded by the right position in the reactor, and a new HDM catalyst (HDM-3), HDM/HDS catalyst (HDS-4), and the above highly selective denitrogenation catalyst (HDS-5) suitable for DAO treatment could be developed. Thus, it was possible to determine the required catalyst volume (**Fig. 5**).

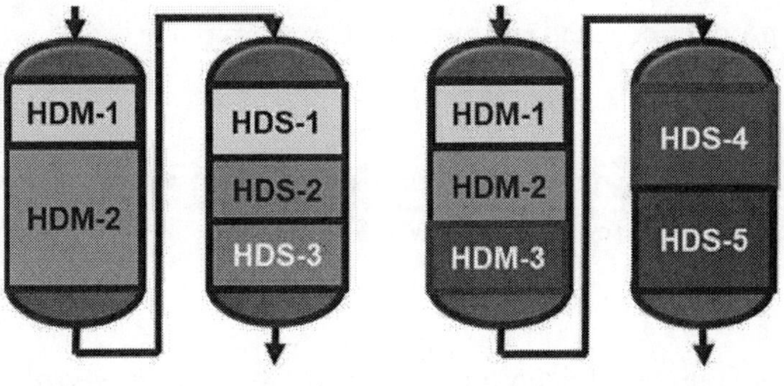

Figure 5. Conventional catalyst system (left) and developed catalyst system (right)

5. RDS commercial operation and RFCC performance improvement

After verifying the developed catalyst system through an aging test in our pilot plant, commercial operation began in June 2019 in our RDS unit. By applying the developed catalyst system, it became possible to process heavier feedstock with a higher sulfur content than before, while maintaining a high desulfurization rate and stable operation for one year (**Table 1**). Furthermore, the significant reduction of the nitrogen concentration in the RDS product oil, which is a feedstock of RFCC, improved the conversion and gasoline yield in the RFCC unit. It has also become possible to co-produce low sulfur fuel oil that complies with IMO regulations.

Table 1. Effect of developed catalyst system on RDS and RFCC units in refinery

Typical value		Conventional catalyst system	Developed catalyst system
Sulfur content in RDS feed	wt%	Base	+1.0
RDS reactivity			
HDS conversion	%	Base	+4
HDM conversion	%	Base	+3
Nitrogen content in product oil	wtppm	Base	▼660
RFCC reactivity			
Conversion	%	Base	+1.9
Gasoline yield	%	Base	+1.6

6. Conclusions

A residue upgrading technique specializing for the DAO processing was developed. DAO is obtained by SDA processing the VR fraction, which is expected to be surplus in the future. Detailed analysis of the DAO, countermeasures against incompatibility when AR and DAO are mixed, development of effective desulfurization catalyst for DAO processing, and development of simulators for a reaction prediction and a pressure drop prediction were achieved, while heavier crude oil than before was processed. We have established a technique that makes it possible to reduce the sulfur content of RDS product oil and improve RFCC performance.

Acknowledgements

This research was conducted as part of a project subsidized by the Ministry of Economy, Trade and Industry of Japan (METI).

References

1. Matsushita, K., Goto, Y., Takamura, T., Iwama, M., *J. Jpn. petrol, Inst.*, **63**, (5), 308 (2020).

A high-throughput approach to study cation-exchanged zeolites

Hassan Aljama
R&DC, Saudi Aramco

Zeolites are nowadays abundantly used as adsorbents and catalysts for many industrial applications. Computational studies have recently complemented the widely available experimental data and observations on zeolites by shedding light on reaction mechanisms and the active sites. The computational studies of zeolites, however, are challenging due to the numerous possible active sites. A zeolite structure can have many unique tetrahedral sites (T). There are numerous possibilities for the placement of Al and the compensating proton or cation. The number of possibilities grow exponentially for higher Al/Si ratios in the framework. Due to this complexity, most theoretical studies have focused on a limited number of T sites based on physical intuitions. These choices, however, can miss promising active sites. In this work, we attempt to thoroughly explore the possible active sites using a high-throughput approach. We apply our approach to the problem of NO adsorption on metal-exchanged zeolites. It is known that Pd-exchange CHA, for example, is an effective adsorbent to limit NOx emissions. Full exploration of this problem requires enumerating the possible Al-Al pairs on the starting zeolites, identifying the favorable sites for the compensating protons and cations and searching for a global minimum in each case, which means thousands of calculations even for simple zeolites. We show how we systematically explore the possible sites. The automation of this calculation reduces the amount of manual labor and highly improve the efficiency of the workflow. It presents a new approach to tackle the complicated problems related to zeolite catalysis.

Catalytic manufacture of ethylene–1-hexene thermoplastic elastomer: Thermodynamic process development

M. Atiqullah*[1], A.F.A. Abdelaal[2], S. Adamu[3], and H. S. Alasiri[1, 3]
*Presenter, Senior Research Engineer (Professor), Email: matiq@kfupm.edu.sa
[1]Interdisciplinary Research Center for Refining & Advanced Chemicals (IRCRAC), Research Institute
[2]Department of Mechanical Engineering, [3]Department of Chemical Engineering
King Fahd University of Petroleum & Minerals, Dhahran, Saudi Arabia

Extended Abstract

Diversification and enhancement of non-oil revenue make a major objective of Vision 2030 of Saudi Arabia. In this context, conversion of local feedstock into value-added products, which can expand Saudi economy and employment, is highly imperative. Saudi Arabia's current alpha-olefin production is 150 kmtpa which will increase to 550 kmpta by 2025. Hence, IRCRAC's alpha-olefin valorization research focuses on upgrading the utilization of this raw material thru selectively making thermoplastic elastomers (TPEs), polymeric flow-aids, and synthetic lubricants. The present talk exclusively confines to ethylene-1-hexene thermoplastic elastomer (E-1H TPE) production. Saudi Arabia is the 2nd largest global polyolefin producer (16.25 mmpta) while the worldwide TPE production will rise to 7.38 mmpta by 2022. Applications of TPEs include automotive, building and construction, footwear, appliances, house wares, medical and hygiene, wire and cable, sport/leisure/toys, packaging, etc. E-1H TPEs can expand the applications of thermoplastic as well as specialty polyolefins. Polyolefin blends and composites occupy a large part of them. Such wide application profile and market also hold strong potential for establishing medium-size start-ups, which befit Vision 2030. Accordingly, the present talk is designed into the following four parts.

Catalyst is the heart of E-1H TPE manufacture process which can be slurry or solution. In either situation, ethylene solubility in the solvent–1-hexene binary mixture is a key process parameter. Industrial solvents include toluene and *n*-hexane. Hence, **Part I** underscores ethylene solubility and novel catalyst development. Ethylene participates in copolymerization initiation and propagation steps (as per the original Cossee-Arlman migratory olefin insertion mechanism and its subsequent refinements). See Figure 1. Consequently, ethylene solubility influences polymerization mechanism and kinetics, catalyst activity and the resulting copolymer molecular weight, which in turn affect end-product performance. Precatalysts such as Spaleck and constrained geometry types, tridentate titanium and tetradentate zirconium post-metallocenes, and Nomura (cyclopentadienyl)(aryloxy)titanium post-metallocenes show promise in the laboratory to incorporate high level of 1-hexene. But to transform these potential laboratory results into industrial applications is a challenge. Hence, **Part I** particularly focuses on this. Figure 2 illustrates how increasing incorporation of 1-hexene transforms a semicrystalline copolymer into the TPE analogue.

Part II addresses literature solubility model and algorithm limitations that list (i) convergence difficulty, (ii) inapplicability particularly to high industrial temperature and pressure, (iii) inadequacy to explain solvent structure, temperature, and pressure effect on solubility, and (iv) inability to illustrate phase instability.

Part III, therefore, reports how the present model, developed using thermodynamic vapor-liquid equilibrium (VLE) concept, Peng-Robinson equation of state, and fluid thermophysical properties, differs from the existing ones and overcome the above limitations. The above thermodynamic model framework is illustrated in Figure 1. We do not any further need to a priori assume, iterate, or calculate equilibrium constant (*K* factor), phase fraction, bubble point, dew point, and liquid phase activity coefficients. The model predictions well match experimental data and present a robust and efficient algorithm that applies to laboratory to industrial polymerization conditions (0.1–10 MPa and RT–200 °C). See Figure 3. We also show the effects of liquid phase fluid compressibility and ethylene-solvent interaction on solubility, catalyst active species behavior in solvent environment, monomer insertion, and copolymer chain growth. Figure 4 demonstrates these features. Ethylene solubility in an asymmetric binary mixture of toluene–1-hexene and *n*-hexane–1-hexene, despite being so important, remains completely unaddressed. Note that solvent–comonomer-rich feed makes high 1-hexene-containing E-1H copolymers. But customarily, ethylene solubility in a single solvent such as toluene or *n*-hexane is used to design the related copolymerization trials, which is conceptually incorrect. The present model remedies this. See Figure 5.

Part IV shows that now we can much better investigate the related catalyst design and synthesis, causes of catalyst deactivation, polymerization mechanism and kinetics, monomer saturation behavior, fluid compressibility behavior, Henry's constant, and entropy departure effect on ethylene solubility. Figures 4 and 6 summarize these phenomena. This is how the present work adds new insight and provides missing explanations.

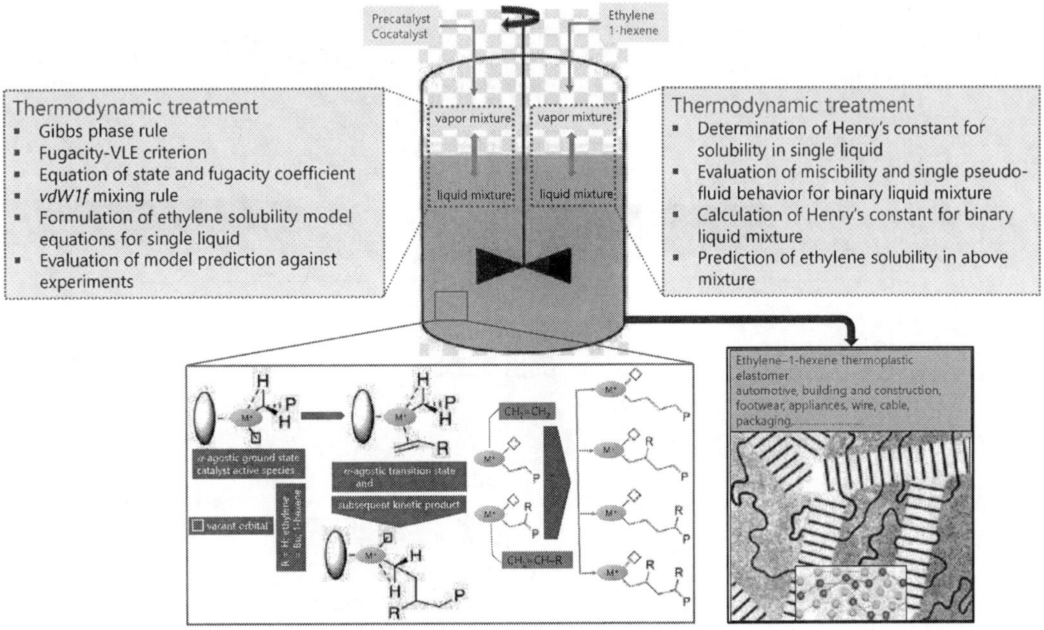

Figure 1. Present study research theme.

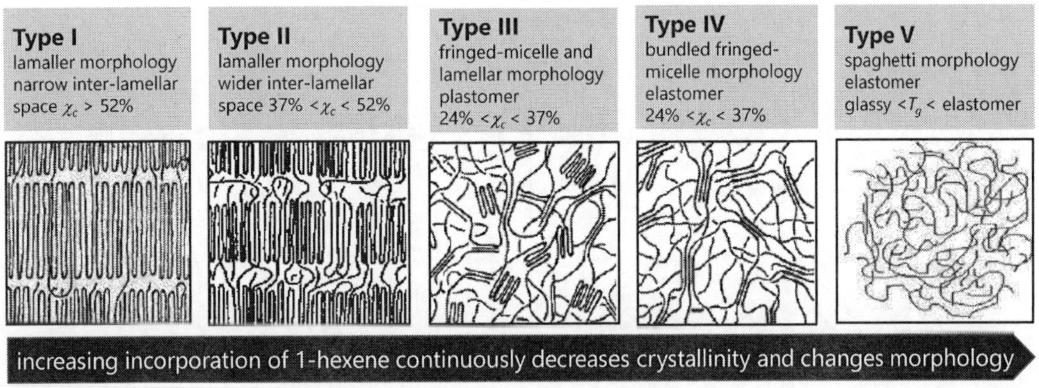

Figure 2. Ethylene–1-hexene copolymer morphology progression.

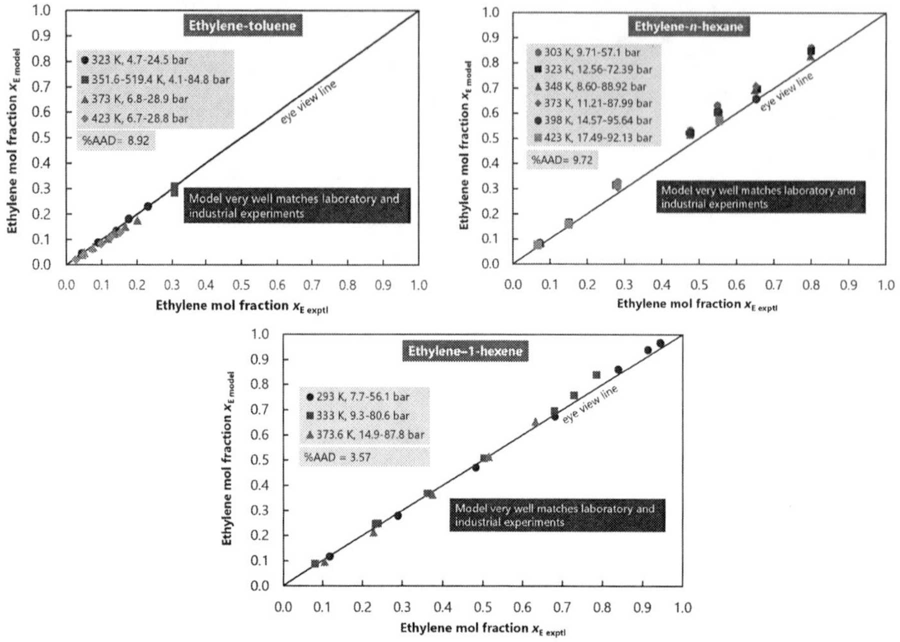

Figure 3. Solubility of ethylene in toluene, *n*-hexane, and 1-hexene: model versus experiment over laboratory to industrial conditions.

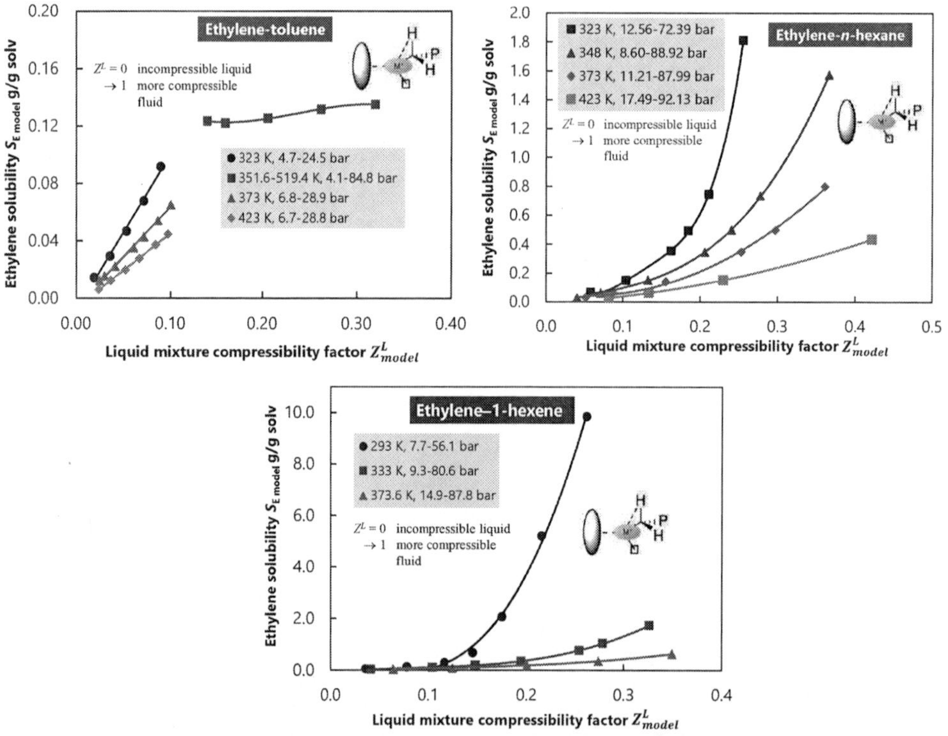

Figure 4. Effect of liquid phase fluid compressibility and ethylene-solvent interaction on solubility, catalyst active species, monomer insertion, and copolymer chain growth.

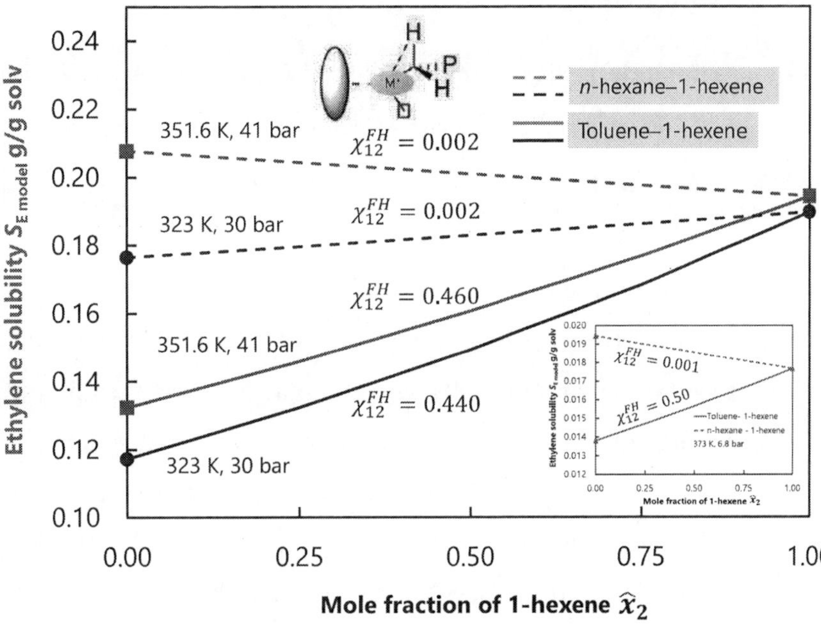

Figure 5. Model-predicted solubility of ethylene in toluene–1-hexene and *n*-hexane–1-hexene binary solvents.

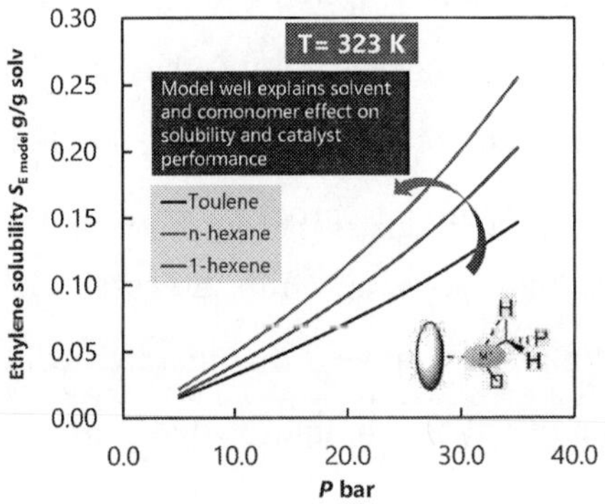

Figure 6a. Variation of ethylene solubility as a function of solvent type.

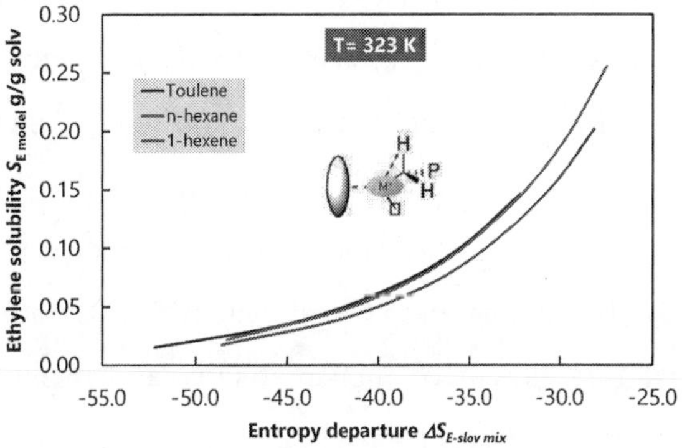

Figure 6b. Effect of entropy departure on ethylene solubility as a function of solvent type.

Hierarchical Nano-sized Zeolite Beta Development for Crude Oil Hydrocracking Catalysts

Lianhui Ding, Faisal Alotaibi, Husin Sitepu, Anaam Al-ShaikhAli, Hana Habboubi, Mohammad Aljishi

Research and Development Center, Saudi Aramco

Abstract

Hydrocracking is one of the most important heavy oil conversion processes in modern refineries and petrochemicals plants. Hydrocracking catalysts play a critical role in determining the product slates and properties, environmentally safe and thus it is benefites from the economics point of view. Zeolites are cracking component of the commercialized hydrocracking catalysts. Its pore structure and acidity can be modified by various chemical or steaming treatments, and have critical effect on the catalyst reaction performances. In present study, the hierarchical nano-sized zeolite beta was synthesized by desilication and hydrothermal treatment of the non-calcinated and calcinated nano-sized zeolite beta. The mesopores are not only generated via intra-particle voids, but also inside the zeolites. Characterizations of the high-resolution X-ray powder diffraction (XRD) data, the Brunauer-Emmett-Teller (BET), and the ^{27}Al ssolid-state nuclear magnetic resonance (NMR) spectroscopy of the samples show that the zeolite crystallinity is well preserved for non-calcinated zeolite due to the fact that the template protection is either in desilication or hydrothermal treatment process. The reaction performances in the arab light (AL) crude oil hydrocracking were carefully assessed with the modified zeolite based NiMo catalysts in pilot plant. The results showed that hierarchical nano-sized zeolites have higher activity and naphtha selectivity than commercialized micro-sized zeolite beta-based catalysts.

Digital Chemistry Guided Development of Novel Ligands for Selective Ethylene Tetramerization

Sibo Lin[1], Mohamed Elanany[2], Motaz Khawaji[2], Wei Xu[2], E. A. Jasser[3], Samir Barman[3] and Nestor Garcia[3]

[1]Aramco Services Company: Aramco Research Center – Boston, Catalysis Technology Group, Cambridge, MA, USA, siboster@gmail.com
[2] Saudi Aramco: Aramco Research Center – KAUST, Fuels and Chemicals R&D Division, Thuwal, Saudi Arabia
[3] Center for Refining and Advanced Chemicals, King Fahd University of Petroleum and Minerals, Dhahran 31261, Saudi Arabia

Abstract:

Oligomerization of ethylene to linear alpha olefins is an important industrial catalytic process. Conventional non-selective catalysts produce a broad Schulz-Flory distribution of various ethylene oligomers, many of which are low-value. To target on-demand production of valuable 1-octene, highly active homogeneous catalysts based on Cr active sites and supported by diphosphinoamine (PNP) ligands have been known to selectively tetramerize ethylene for nearly two decades. However, deployment of these catalysts is hindered economically by inevitable formation of byproducts including 1-hexene, methylcyclopentane, methylenecyclopentane, and polyethylene, which is especially pernicious as its insolubility and build-up in the reactor leads to process fouling and shutdown. To accelerate catalyst discovery, we have developed several digital chemistry strategies to help identify promising PNP variants. (1) PNP ligand solubility is correlated with catalyst activity. (2) Calculated thermodynamic stability of PNP ligand against isomerization to bisphosphinoimine (PPN), calculated via density functional theory (DFT), is correlated with lower PE selectivity. (3) The steric bulk of the ligand, measured via SambVca buried volume calculations, is correlated with various product selectivities. Combining these various digital chemistry guides, we have developed and patented novel PNP ligands with higher catalytic activity, lower PE selectivity, and greater overall linear alpha olefin selectivity.

High-Throughput Screening and Literature Data Driven Machine Learning Assisting Discovery of La_2O_3-based Catalysts for Low-Temperature Oxidative Coupling of Methane

Shun Nishimura

Graduate School of Advanced Science and Technology, Japan Advanced Institute of Science and Technology (JAIST), 1-1 Asahidai, Nomi, Ishikawa, 923-1292, Japan

E-mail: s_nishim@jaist.ac.jp

Abstract

Three-element component La_2O_3-based catalysts for the oxidative coupling of methane (OCM) are examined for this study using machine learning (ML) approaches such as support vector regression (SVR) and random forest regression (RFR). This validation was conducted while assuming that the three-element component (M1–M2–M3) resulting in high C_2 yield predicted by ML is helpful to ascertain the appropriate component to promote the unique nature of La_2O_3 itself: the low-temperature OCM feature. The combined use of an open-source high-throughput screening (HTS) database and SVR assisted the discovery of three-element component La_2O_3-based OCM catalysts of 11 types with C_2 yield (> 5.0%) appearing at 450°C. Then, to predict more unique component La_2O_3-based OCM catalysts from the outer field of screening elements at HTS experimentation, HTS and literature databases were applied for SVR and RFR. This combined approach discovered 11 additional combinations of three-element component La_2O_3-based OCM catalysts affording C_2 yield (> 5.0%) appearing at 450°C.

1. Introduction

Because of the increasing availability of natural gas resources such as shale gas and methane hydrate, catalytic transformation of methane has attracted strong attention in industrial and academic fields. The oxidative coupling of methane (OCM) discovered in the 1980s is one attractive process for producing C_2 hydrocarbons of ethylene (C_2H_4) and ethane (C_2H_6) directly from methane (CH_4) [1-2]. However, using conventional processes of catalyst discovery based on scientists' intuition and experience, the desired performance criteria of a C_2 yield higher than 30% and superior C_2 selectivity of 80% with an appropriate C_2 yield sufficient to generate economically rational industrial plant systems have persisted as great challenges for the past 40 years [3-6].

Along with the recent intense growth of data science management and machine learning (ML) engineering, a common interesting query has arisen: "How can such new technologies aid catalyst discovery more effectively in comparison with conventional method?". *Catalyst Informatics* has been adopted to develop next-generation understanding as a trend of data used for catalyst discovery [7-10]. Indeed, some successful discoveries of homogeneous catalysts have been achieved [11-12]. Such discoveries rely on the nature of homogeneous catalysis, which is fundamentally attributable to the design of an active metal center and a coordinating ligand. When applying a heterogeneous (solid) catalyst, greatly challenging issues arise for application because heterogeneous catalysis depends on numerous catalyst features such as composition, morphology (size, shape, crystallinity, etc), oxidation state, defect, strain, support nature, additive element, and stability under active conditions. In fact, ascertaining the relations among physicochemical features as determined by their reactivity and by conventional characterizations is extremely complicated. Moreover, even if some clear relation is identified between those characteristic features and reactivity, it remains controversial whether the relation can serve in a common role as a global descriptor. In other words, the model of representing heterogeneous catalysis feature using ML algorithm networking is an extremely important and persistent subject [13].

Since 2017, the author has explored how to apply the ML engineering to produce benefit information for catalyst scientists under close collaboration with ML engineers and data scientists, as presented later in Acknowledgements. During initial stages of research, results demonstrated that a well-defined random forest regression (RFR) based on a large dataset from the literature including 1,868 catalyst data can propose 56 undiscovered catalyst components and corresponding reaction conditions for OCM to achieve C_2 yield higher than 30% [14]. It is particularly interesting that the RFR also revealed the hidden key importance of OCM catalysis among the 11 descriptors of the preparation method, reaction temperature, CH_4 pressure, O_2 pressure, contact time, cation1, cation1 mol%, cation2, cation2 mol%, catalyst support, and catalyst support mol%. Those findings indicated ML approaches as effective tools for discovering unreported catalysts and for elucidating hidden effects such as catalyst components and reaction conditions on catalyst performance. However, at the next research stage in validation, a difficulty arose in achieving target OCM catalyst performance with a high C_2 yield over 30% among the predicted catalyst components [15]. One more impressive experience was conducted on tracing the catalyst behaviors under different reaction conditions by interpolate filling using ML engineering with a small experimental dataset. At this time, the author examined the OCM reaction over a well-known Na-Mn-W/SiO_2 catalyst for 156 data points collected one-by-one using a conventional fixed-bed reactor for experimentation. Then, ML was implemented to treat experimental conditions in high dimensions described with five reaction parameters: reaction temperature, CH_4/O_2 ratio, CH_4 and O_2 gas concentrations (CH_4 + O_2 conc.), total flow rate, and catalyst weight. At this time, extra tree regression (ETR) was found to represent the trend of experimental data accurately [16]. This representation predicted that the target of C_2 yield features over Na-Mn-W/SiO_2 can help to indicate the best reaction conditions to afford the best performance. Therefore, ML was demonstrated as a powerful approach for ascertaining and then elucidating the best experimental conditions in multiple dimensions. Other earlier studies of reaction network representation [17-18] and hidden rule identification by item-set mining [19] have also provided attractive findings for improvement of scientists' understanding of some catalysis features. According to these shared experiences with collaborators, the author has found that ML and data management engineering skills hold great potential to reveal hidden trends of data in both large and small datasets, but exact prediction of the outer area performance is apparently still challenging. It is worth describing that the former is apparently difficult by human senses, but the latter has been achieved by conventional catalyst scientists without such a complete understanding, in catalyst development and modification history. Therefore, arranging the ML-aided catalyst discovery with more appropriate methods for generating unexpectable components possessing high performance at target reactions is truly desired.

Along with the intense growth of data analysis and predictions that have been made using ML and data management engineering tools, it has become necessary to collect huge amounts of tailor-made datasets with high efficiency. Several parallel reactor systems have been developed in a fixed bed catalytic reactor [20-23]. The combined use of tailor-made datasets for experimentation that have complied using systematic high-throughput screening (HTS) experimentation and ML and data management engineering skills have opened up different protocols for understanding catalysis features at catalyst discovery [24-29]. One earlier study used OCM catalyst discovery from a bias-free and consistent process for HTS at OCM. They described their approach as "Catalyst discovery from Scratch" [30]. For this protocol, three active elements (M1–M2–M3) were selected randomly from Li, Na, Mg, K, Ca, Ti, V, Mn, Fe, Co, Ni, Cu, Zn, Sr, Y, Zr, Mo, Pd, Cs, Ba, La, Ce, Nd, Eu, Tb, Hf, W, or "none" to make 28 selections. Then, these were co-impregnated onto the support, which was also selected randomly from MgO, gamma-Al_2O_3, SiO_2, $Ca(OH)_2$, anatase-TiO_2, ZrO_2, $Ba(OH)_2$, La_2O_3, or CeO_2 to make nine selections. The dried powder was calcined at 1000°C for 3 h to supply mixed-oxide form. Then totally 300 catalysts were prepared and checked for reactivity under various reaction condition using the systematic HTS system. The generated bias-free catalyst library and data analysis with visualization, ML engineering, and constructed networking provided a perspective for views on OCM catalysis features that were different from those assisted with conventional catalyst chemistry methods [13, 31-33]. It is proposed that those studies based on systematic HTS may alter the style of catalyst discovery. However, the utility of such powerful tools of HTS has limited their popularity because of associated costs for construction and operation at present situation. Nowadays, the databases also can be accessible from the literature and/or an on-line open platform such as CADS [34-35]. Therefore, a need exists for discussing a means of applying open-resource datasets collected by different researchers including HTS datasets to assist various researchers during catalyst discovery and modification.

Herein, the author specifically examines the low-temperature OCM character derived La_2O_3 catalyst [19]. Reportedly, lanthanum-based catalysts such as La_2O_3 nanorod [36], $La_2O_3CO_3$ nanorods [37], and $La_2Ce_2O_3$ perovskite [38-39] have shown especially low onset temperatures of OCM: 420-550°C. However, multi-element-supported La_2O_3 features have rarely been investigated [22]. Therefore, the author believes that ML-aided catalyst investigation revealing a new multi-element-supported La_2O_3 catalyst possessing much unique lower temperature OCM feature has become a notable achievement [40].

2. Experimental

Three-element-supported La_2O_3 catalysts were prepared using co-impregnation method with an aqueous solvent of mixed metal resources at 50°C for 6 h. The loading amount for each metal was fixed at 0.3 wt%, in theory. After evaporating the water solvent, the remaining mixed powder was dried at 110°C. Then it was calcined at 600°C for 3 h. No-metal-supported La_2O_3 (bare), denoted as none/La_2O_3, was treated using the same protocol. The OCM reaction was performed with as-prepared catalyst (50 mg) in a conventional fixed-bed reactor system (downstream) with a furnace (L = 270 mm) in 50°C intervals at 400-850°C. Pre-treatment was applied at 400°C for 1 h under an O_2 flow (20.0 ml min^{-1}). Then, a typical OCM stream composed of $CH_4/O_2/N_2$ = 12.0/6.0/3.0 ml min^{-1} was used to evaluate the OCM reactivity. It is noteworthy that a standard Na-Mn-W/SiO_2 gave a significant C_2 yield of 24.8% at 800°C, whereas none/La_2O_3 served C_2 yield of 11.2% at 500°C

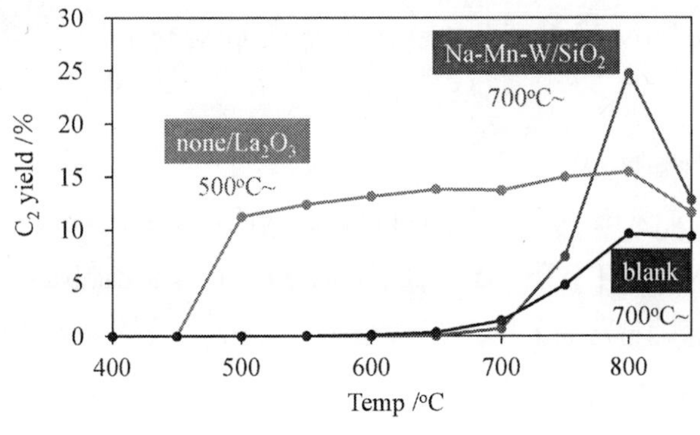

Figure 1. Trends of typical OCM catalyst at present reaction condition at CH_4/O_2 = 2.0.

under the conditions ($CH_4/O_2 = 2.0$) shown in **Figure 1**. The findings indicated clearly that the former had a high potential at higher temperature. The latter possessed a unique feature as a lower onset temperature in OCM reaction.

Then, ML prediction was conducted using the open-source HTS data (40,330 points by 350 catalysts) collected by a collaborator. The data were from surveys of OCM catalyst potential at 700–1000°C under various reaction conditions of the CH_4/O_2 ratio, total flow volume, and $CH_4 + O_2$ conc. [20, 30]. To describe the catalyst composite information of three-element-impregnated catalyst (M1–M2–M3/Support), a one-hot encoding manner was used. Binary numbers 1 and 0 were assigned to the corresponding index cell for composite element and support. The scheme was implemented in accordance with reaction conditions and the corresponding C_2 yield (**Figure 2**). Furthermore, for extracting appropriate trends of data, the data for validation were selected in accordance with the following conditions: total flow greater than 20 mL min^{-1}, $CH_4 + O_2$ conc. over 85%, and $CH_4/O_2 = 2.0$. A difficult issue exists for HTS dataset-driven ML at this study: because the HTS data were obtained for higher temperatures (700–1000°C) but not for lower temperature trends (approx. 400–600°C) of OCM performance, direct validation of predicted C_2 yield might have insufficient guidance to discover low-temperature La_2O_3-based OCM catalysts. To overcome this shortcoming, the author revisited the research history made with collaborators using ML engineering tools (*vide spuria*) for interpretation of the meanings of the trends covered by ML regression models. Consequently, the assumption that the ML regression field might emphasize friendly element combinations (M1–M2–M3) for OCM showing higher C_2 yield was applied for this study to ascertain components that are useful for enhancing the unique characteristics for La_2O_3.

	[Cat. comp. described with a one-hot encoding manner]								[Reaction condition]				[Target]
Cat. Comp.	Ba	Ca	Mn	Co	Na	Al_2O_3	SiO_2	· · · · ·	Temp.	p_{CH4}	p_{O2}	$p_{balance}$	C_2 yield
$BaCaMn/Al_2O_3$:	1	1	1	0	0	1	0	· · · · ·	1273 K	0.56	0.29	0.15	x %
$CaMnCo/SiO_2$:	0	1	1	1	0	0	1	· · · · ·	973 K				y %
$MnCoNa/SiO_2$:	0	0	1	1	1	0	1	· · · · ·	1173 K				z %

Figure 2. Example of one-hot encoding implementation at this study.

3. Results and Discussion

First, a support vector regression (SVR) model was applied to represent C_2 yields at 40,330 points (350 catalysts) covered by HTS. The data points derived from nine catalysts were removed because of extraordinary trends in selectivity. This model score at cross-validation of train-and-test examination was 0.86. Then, specifically examining the M1–M2–M3/La_2O_3 category at the constructed SVR field predicted 41 three-element combinations at La_2O_3-based catalysts with features affording C_2 yields higher than 16.01% at higher temperatures. The TOP 20 were selected from the SVR field at La_2O_3 support. Then their potential for use as a low-temperature OCM based on La_2O_3 was examined, as shown in **Figure 3**. The expected score of C_2 yield at the SVR field, the predicted C_2 vale was higher to the left side in the figure, did not contribute directly to the enhancement capacity of La_2O_3 character over that of low onset temperature. However, an interesting finding was that M1–M2–M3 components of 11 types can afford C_2 yield at 450°C, which was a lower onset temperature than the 500°C observed for none/La_2O_3. The components (C_2 yield at 450°C) were YEuHf (13.1%), CaYHf (10.1%), CaYEu (15.1%), CaYLa (15.0%), CaYBa (13.6%), MgNdHf (9.5%), CaEuHf (13.2%), CaYNd (10.2%), SrNdHf (9.7%), CaNiY (6.1%), and MgEuHf (11.4%). Although the remaining nine validations were not matched with the assumption, one was more likely to find undiscovered and effective catalysts with such *indirect* ML effectively from different areas of the collected datasets. Indeed, the hit rate for finding the low-temperature OCM catalysts was 55% (11 appearances per 20 validations), which is apparently a highly attractive rate.

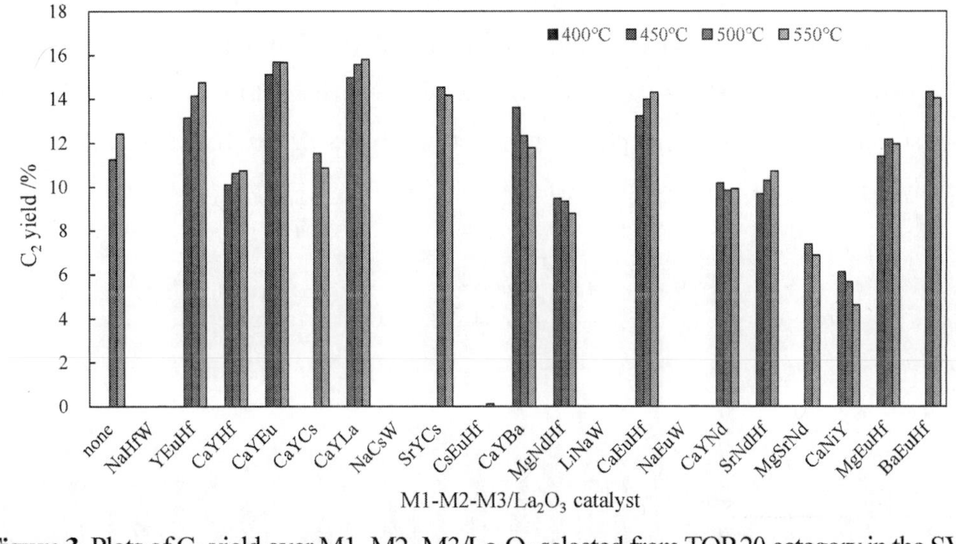

Figure 3. Plots of C_2 yield over M1–M2–M3/La_2O_3 selected from TOP 20 category in the SVR field based on HTS data. The order of element from left to right was corresponding to the predicted C_2 yield from high to low values at the SVR field.[10, 40]

Next, implementation of data from the literature combined with the tailor-made HTS data was examined. The HTS data were compiled by 28 selections of elements and 9 selections of supports, as described above. Therefore, the output area of ML prediction using a one-hot encoding method was limited within such original indexes. To investigate more undiscovered La_2O_3-based catalyst, data from the literature of 1,802 points (1,286 catalysts) [3] were added for subsequent ML prediction approaches. At that time, 59 catalysts derived from Na-Mn-W based catalyst in HTS [20] were excluded. Also, to eliminate data biases derived from the Na-Mn-W-like category and various reaction conditions at HTS experimentations, the best C_2 yield values were applied for constructing an ML regression model from 300 random HTS data [30]. Then, the score at renewed SVR became lower to be 0.45 at the train-and-test examination. That might be attributed to bias in the literature data, which were inferred from different reactor systems and conditions at variety laboratories. However, when this challenge implies a different possible area, it would be an interesting strategy for three-element component enhancing the La_2O_3 nature as a low onset temperature OCM catalyst. Random forest regression (RFR) was also applied to the processed data: its score was 0.51. The individual TOP 12 of M1–M2–M3/La_2O_3 components selected from SVR and RFR fields at the predicted C_2 yield were prepared and examined for those features at OCM. It is particularly interesting that MgSrYb (C_2 yield at 450°C: 12.1%) and GaSrYb (14.9%) from SVR assistance and GaTbHf (7.5%), ZnGaHf (3.8%), GaEuHf (7.5%), GaYEu (5.5%), ZnGaY (7.2%), ZnGaEu (9.3%), GaYbHf (6.3%), GaSrEu (7.6%), GaSrY (8.3%), GaSrHf (1.8%), and GaEuTb (8.6%) from RFR assistance were newly identified as low-temperature OCM catalysts based on La_2O_3 in each of 12 validations, as shown in **Figure 4**. Only the MgSrNd duplicated earlier experimentation by SVR using HTS data (**Figure 3**), although this was inactive at 450°C. Although RFR was composed mainly by GaHf derivatives, results showed that the addition of literature data into the HTS datasets can be an effective approach for emphasizing the outer area capacity in ML prediction. An additional 11 catalysts possessed C_2 yields higher than 5.0% at 450°C.

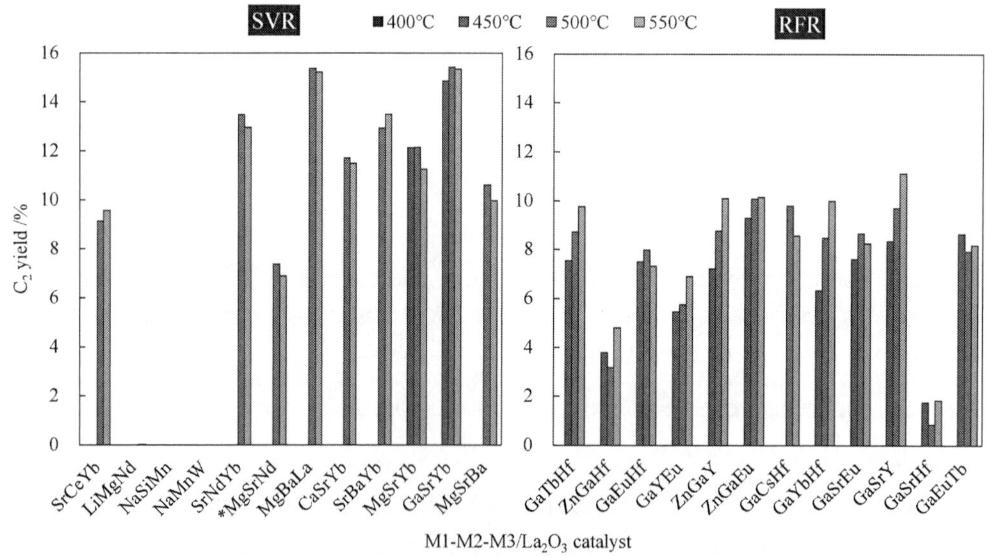

Figure 4. Plots of C_2 yield over M1–M2–M3/La_2O_3 selected from TOP 12 category in the SVR and RFR field based on HTS and literature data. The order of element from left to right corresponding to the predicted C_2 yield from high to low values, respectively, at SVR and RFR field.[40]

To emphasized the important potential of 24 as-revealed La_2O_3-based catalysts, the selected 10 catalysts affording C_2 yields higher than 10.0% at 450°C under $CH_4/O_2 = 2.0$ were examined further to assess the OCM performance at $CH_4/O_2 = 3.5$, which condition is a stricter condition for CH_4 activation at low temperatures. Under the present conditions, none/La_2O_3 showed an onset temperature of 600°C for appearance of C_2 yield (1.2%). **Figure 4** shows that all components retained a positive effect on the La_2O_3 feature of a low onset temperature in OCM at $CH_4/O_2 = 3.5$. It is noteworthy that the two catalysts of CaYHf and MgEuHf exhibited C_2 yields at 450°C of 3.6% and 3.1%, respectively, even under the $CH_4/O_2 = 3.5$ condition. Additionally, the two catalysts of CaYBa and YEuHf respectively showed attractive C_2 yield of 8.3% and 2.8% at 500°C. These results indicate that the as-revealed La_2O_3 catalysts promoted positive potential for further investigations, such as performance improvement and mechanistic study at next stage in conventional methods used for catalyst chemistry.

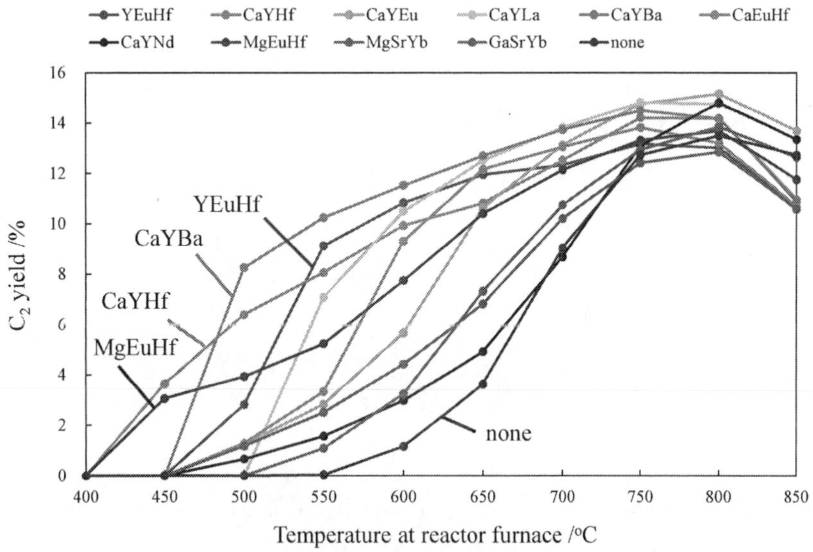

Figure 4. Plots of C_2 yield as a function of reactor temperature over selected 11 types of M1–M2–M3/La_2O_3 at $CH_4/O_2 = 3.5$.[40]

4. Conclusions

Specifically examining the unique OCM features derived from La_2O_3 catalyst, this investigation of multiple components supported La_2O_3 catalyst was conducted by ML engineering for the combined use of open-resource HTS datasets collected by a collaborator [20, 30] and literature data summarized in the review article [3]. To overcome mismatching between the target temperature area (approx. 400–600°C) and the collected data area (mainly at around 700–1000°C) in the dataset resources, *indirect* ML prediction was used. Investigation was done based on the hypothesis that the M1–M2–M3 combinations giving a high C_2 yield at high temperature predicted by a direct ML prediction are helpful for finding familiar multiple components on the La_2O_3 support, consequently enhancing the unique nature of low onset temperature in OCM for La_2O_3 itself. It is particularly interesting that three elements of 11 types in 20 validations were found to be effective M1–M2–M3 components when using HTS data-driven ML with SVR. Moreover, the combined use of HTS and data from the literature further revealed additional 11-13 types of active M1–M2–M3 component La_2O_3-based catalysts at 450°C. This finding can promote further studies of improvement, further revealing the roles of the respective components at next step in conventional styles of catalyst chemistry. Accordingly, this report presents a successfully applied approach for discovering heterogeneous catalysts with ML assistance [40]. The use of ML to identify knowledge from trends of data to supply new ideas for conventional scientists has also been presented in reports of our recent studies [41-43].

Acknowledgement

This work was conducted with Prof. K. Takahashi (Hokkaido Univ.), Prof. J. Ohyama (Kumamoto Univ.), Prof. T. Taniike (JAIST), and Prof. T. Uno (NII), under financial support by JST-CREST (Grant No. JPMJCR17P2) in Japan. Prof. K. Takahashi is the team reader for our Catalyst Informatics project started in 2017 (see: https://catalystinfo.eng.hokudai.ac.jp/index-e.html).

References

1. G. E. Keller, M. M. Bhasin, J. Catal., 73, 9 (1982).
2. W. Hinsen, M. Baerns, Chem.-Ztg., 107, 223 (1983).
3. U. Zavyalova, M. Holena, R. Schlogl, M. Baerns, ChemCatChem, 3, 1935 (2011).
4. E. V. Kondratenko, T. Peppel, D. Seeburg, V. A. Kondratenko, N. Kalevaru, A. Martin, S. Wohlrab, Catal. Sci. Technol., 7, 366 (2017).
5. R. Schmack, A. Friedrich, E. V. Kondratenko, J. Polte, A. Werwatz, R. Kraehnert, Nat. Commun., 10, 441 (2019).
6. S. Mine, M. Takao, T. Yamaguchi, T. Toyao, Z. Maeno, S. M. A. H. Siddiki, S. Takakusagi, K. Shimizu, I. Takigawa, ChemCatChem, 13, 3636 (2021).
7. K. Takahashi, L. Takahashi, I. Miyazato, J. Fujima, Y. Tanaka, T. Uno, H. Satoh, K. Ohno, M. Nishida, K. Hirai, J. Ohyama, T. N. Nguyen, S. Nishimura, T. Taniike, ChemCatChem, 11, 1146 (2009).
8. T. Toyao, Z. Maeno, S. Takakusagi, T. Kamachi, I. Takigawa, K. Shimizu, ACS Catal., 10, 2260

(2020).
9. K. Takahashi, L. Takahashi, S. D. Le, T. Kinoshita, S. Nishimura, J. Ohyama, J. Am. Chem. Soc., 144, 15735 (2022).
10. K. Takahashi, L. Takahashi, S. Nishimura, J. Fujima, J. Ohyama, Chapter 12: The Rise of Catalysts Infomatics, in "Crystalline Metal Oxide Catalysts" by W. Ueda (ed.), Springer (2022).
11. M. S. Sigman, K. C. Harpe, E. N. Bess, A. Milo, Acc Chem Res., 49, 1292 (2016).
12. M. Mukai, K. Nagao, S. Yamaguchi, H. Ohmiya, Bull. Chem. Soc. Jpn., 95, 271 (2022).
13. S. Ishioka, A. Fujiwara, S. Nakanowatari, L. Takahashi, T. Taniike, K. Takahashi, ACS Catal. 12, 11541 (2022).
14. K. Takahashi, I. Miyazato, S. Nishimura, J. Ohyama, ChemCatChem, 10, 3223 (2018).
15. S. Nishimura, J. Ohyama, T. Kinoshita, S. D. Le, K. Takahashi, ChemCatChem, 12, 5888 (2020).
16. J. Ohyama, S. Nishimura, K. Takahashi, ChemCatChem, 11, 4307 (2019).
17. I. Miyazato, S. Nishimura, L. Takahashi, J. Ohyama, K. Takahashi, J. Phys. Chem. Lett., 11, 787 (2020).
18. L. Takahashi, J. Ohyama, S. Nishimura, K. Takahashi, J. Phys. Chem. Lett., 12, 558 (2021).
19. J. Ohyama, T. Kinoshita, E. Funada, H. Yoshida, M. Machida, S. Nishimura, T. Uno, J. Fujima, I. Miyazato, L. Takahashi, K. Takahashi, Catal. Sci. Technol., 11, 524 (2021).
20. T. N. Nguyen, T. T. P. Nhat, K. Takimoto, A. Thakur, S. Nishimura, J. Ohyama, I. Miyazato, L. Takahashi, J. Fujima, K. Takahashi, T. Taniike, ACS Catal., 10, 921 (2020).
21. C. Ortega, D. Otyuskaya, E. Ras, L. D. Virla, G. S. Patience, H. Dathe, Can. J. Chem, Eng., 99, 1288 (2021).
22. Z. Li, L. He, S. Wang, W. Yi, S. Zou, L. Xiao, J. Fan, ACS Comb. Sci., 19, 15 (2017).
23. C. Hoffmann, H. Schmidt, F. Schuth, J. Catal. 198, 348 (2001).
24. K. Takahashi, L. Takahashi, T. N. Nguyen, A. Thakur, T. Taniike, J. Phys. Chem. Lett., 11 6819 (2020).
25. K. Sugiyama, T. N. Nguyen, S. Nakanowatari, I. Miyazato, T. Taniike, K. Takahashi, ChemCatChem, 13, 952 (2020).
26. I. Miyazato, T. N. Nguyen, L. Takahashi, T. Taniike, K. Takahashi, J. Phys. Chem. Lett., 12, 808 (2021).
27. T. N. Nguyen, K. Seenivasan, S. Nakanowatari, P. Mohan, T. P. N. Tran, S. Nishimura, K. Takahashi, T. Taniike, Mol. Catal., 516, 111976 (2021).
28. Y. Shinke, T. Miyazawa, M. Hiza, I. Nakamura, T. Fujitani, React. Chem. Eng., 6, 1381 (2021).
29. K. Chen, H. Tian, B. Li, S. Angarajan, AIChE J., e17584 (2022).
30. T. N. Nguyen, S. Nakanowatari, T. P. N. Tran, A. Thakur, L. Takahashi, K. Takahashi, T. Taniike,

ACS Catal., 11, 1797 (2021).
31. S. Nakanowatari, T. N. Nguyen, H. Chikuma, A. Fujiwara, K. Seenivasan, A. Thakur, L. Takahashi, K. Takahashi, T. Taniike, ChemCatChem, 13, 3262 (2021).
32. K. Takahashi, J. Fujima, I. Miyazato, S. Nakanowatari, A. Fujiwara, T. N. Nguyen, T. Taniike, L. Takahashi, J. Phys. Chem. Lett., 12, 7335 (2021).
33. L. Takahashi, T. N. Nguyen, S. Nakanowatari, A. Fujiwara, T. Taniike, K. Takahashi, Chem. Sci., 12, 12546 (2021).
34. J. Fujima, Y. Tanaka, I. Miyazato, L. Takahashi, K. Takahashi, React. Chem. Eng., 5, 903 (2020).
35. A web platform Catalyst Acquisition by Data Science (CADS), https://cads.eng.hokudai.ac.jp
36. P. Huang, Y. Zhao, J. Zhang, Y. Zhu, Y. Sun, Nanoscale, 5, 10844 (2013).
37. Y. Hou, W. Han, W. Xia, H. Wan, ACS Catal., 5, 1663 (2015).
38. J. Wu, Y. Zhang, X. Xu, X. Fang, R. Xi, Y. Liu, R. Zheng, X. Wang, ACS Catal., 9, 4030 (2019).
39. Y. Zhang, J. Xu, X. Xu, R. Xi, Y. Liu, X. Fang, X. Wang, Catal. Today, 355, 518 (2020).
40. S. Nishimura, S. D. Le, I. Miyazato, J. Fujima, T. Taniike, J. Ohyama, K. Takahashi, Catal. Sci. Technol., 12, 2766 (2022).
41. S. Nishimura, J. Ohyama, X. Li, I. Miyazato, T. Taniike, K. Takahashi, Ind. Eng. Chem. Res., 61, 8462 (2022).
42. J. Ohyama, A. Hirayama, N. Kondou, H. Yoshida, M. Machida, S. Nishimura, K. Hirai, I. Miyazato, K. Takahashi, Sci. Rep., 11, 2067 (2021).
43. J. Ohyama, Y. Tsuchimura, H. Yoshida, M. Machida, S. Nishimura, K. Takahashi, J. Phys. Chem. C, *Just accepted for publication.*

Optimal design of AI-based models integrated with Ensemble ML paradigms for crude-to-chemicals application

Abdulkadir Tanimu[a], Abdullahi Garba Usman[b,c], Aliyu Musa Alhassan[d], Saheed A. Ganiyu[a,d], Abdullah Aitani[a]

[a] *Center for Refining and Advanced Chemicals, Research Institute, King Fahd University of Petroleum and Minerals, Dhahran 31261, Saudi Arabia*

[b] *Department of Analytical Chemistry, Faculty of Pharmacy, Near East University, TRNC, Mersin 10, 99138 Nicosia, Turkey*

[c] *Operational Research Centre in Healthcare, Near East University, Nicosia, Turkish Republic of Northern Cyprus*

[d] *Department of Chemistry, King Fahd University of Petroleum and Minerals, Dhahran 31261, Saudi Arabia*

Abstract

The dramatic increase in development of high throughput experiments (HTE) in the area of catalysis is attributed to the high amount of data collected within short time. Therefore, the current study utilizes the application of four stand-alone artificial intelligence techniques including; Regression tree (RT), least square BOOST (LSQ-BOOST), Gaussian process regression (GPR) and Robust linear regression (RLR) for modeling the yield of ethylene (C2=) and propylene (C3=) in the catalytic cracking of crude oil feeds using temperature, feed, conversion (%), total gas, dry gas and coke as the independent variables. Correlation matrix and non-linear sensitivity analysis were conducted to filter the input combinations into three different classes (M1, M2 and C3) based on the dependent-independent variables relationship. Four different performance metrics including; Co-efficient of determination (R^2), Pearson correlation co-efficient (PCC), Mean squared error (MSE) and Root mean squared error (RMSE) were used in evaluating the prediction performance of the techniques in both calibration and validations stages. The results of the stand-alone models at certain instances showed weak performance accuracy. Hence, this led to the development of four Ensemble paradigms namely; Generalized Regression Neural Network Ensemble (GRNN-E), Support vector regression ensemble (SVR-E), Feed forward Neural Network Ensemble (NNE) and Adaptive neuro-fuzzy inference system Ensemble in order to improve the performance accuracy of the stand-alone paradigms. The performance efficiency of the Ensemble paradigms depicts that these techniques can improve the performance of the single models up to 58% and 62% in the calibration and validation phases respectively. Lastly, the findings of this study demonstrate the robust application and their possibilities in understanding different catalytic cracking phenomena.

Introduction

Demand for petrochemical feedstocks has been steadily increasing due to the growing demand of petrochemical intermediates that serves as raw materials in the production chain of many industries [1]. The light olefins, particularly ethylene and propylene are important part of these petrochemical feedstocks that have seen increasing demand in recent time [2]. The endothermicity of hydrocarbon cracking is typically responsible for the observed increase in conversion and product yield with temperature [3][4]. Although high temperature result in high conversion and light olefin selectivity, the formation of large amount of dry gas (H_2 and C_1-C_2) and coke under high temperature cracking process is not desirable in refinery processes [5]. High temperature has been linked to monomolecular pyrolytic cracking, where a carbonium ion collapses and preferentially forms H_2, CH_4 and C_2H_6, and a carbenium ion that later transforms to alkene [6]. Thus, increasing reactor temperature for high light olefins yield comes with an expensive choice of producing more dry gas and coke. However, with careful optimization of reactor temperature, an optimum might be reached where the dry gas and coke yield is brought to minimum while ensuring maximum yield of light olefin. Additionally, the type of crude oil feed has been related to the light olefins and dry gas and coke yield in previous studies [3]. Therefore, with careful optimization of crude oil feed and cracking temperature, the yield of light olefins can be maximized while lessening the yield of dry gas and coke. However, this approach involves handling large number of variables that are ordinarily difficult to process using simplified/physical models. The re-emergence of artificial intelligence (AI) in the early 21st century has made analysis and optimization of large number of variables in various academic and industrial application much easier and less laborious [7]. For instance, a comparative study between artificial neural network (ANN) model and nonlinear statistical model showed that the ANN model has higher prediction accuracy [8]. A hybrid neural network model with physical reaction model was discovered to be more efficient in yield optimization and prediction than a conventional reaction model [9]. Recently, Kawai et al.[10] developed an AI hybrid reaction model for the optimization of catalyst make-up rate in a residue fluid catalytic cracking process, which maximize light olefin yield, minimize catalyst loss and offered yield prediction with high level of accuracy.

This study utilizes the application of four stand-alone AI techniques including; Regression tree (RT), least square BOOST (LSQ-BOOST), Gaussian process regression (GPR) and Robust linear regression (RLR) for modeling the yield of ethylene (C2=) and propylene (C3=) in the catalytic cracking of crude oil feeds using temperature, feed, conversion (%), total gas, dry gas and coke as the independent variables. Correlation matrix and non-linear sensitivity analysis were conducted to filter the input combinations into three different classes (M1, M2 and C3) based on the dependent-independent variables relationship. Additionally, four Ensemble paradigms namely; Generalized Regression Neural Network Ensemble (GRNN-E), Support vector regression ensemble (SVR-E), Feed forward Neural Network Ensemble (NNE) and Adaptive neuro-fuzzy inference system Ensemble were developed to further improve the performance accuracy of the stand-alone paradigms.

Catalytic Cracking reaction

The catalytic cracking reaction was carried out according to the ASTM D-7964 method in an Advanced Catalytic Cracking (ACE) unit. The unit is characterized with automated fixed-fluidized bed reactor and was manufactured by Kayser Technology Inc., USA. Three kinds of feeds were cracked using steamed deactivated commercial catalyst at constant CTO ration of 12. Thus, a catalyst weight of 12g was loaded into a catalyst hopper, and feed weight of 1.5 g was injected into the reactor at the rate of 1.2 g/min resulting in a time on stream of 75 s. The reactor temperature was varied between 600 -675 °C while maintaining atmospheric pressure condition.

After the reaction, N2 gas was purged into the reactor for about 9 min for catalyst stripping. The liquid product was collected in a glass receiver that was fixed to the reactor exit and kept in water bath at temperature of −1.25 °C. The total gaseous products weight was measured by water displacement in a gas receiver. After catalyst stripping is finished, the catalyst was regenerated by switching the N2 valve with air and raising the reactor temperature to 700 °C. Thus, all the coke deposited on the catalyst was oxidized to CO2 and analyzed by online IR spectroscopy.

Cracked products analysis

The gaseous products were analyzed online using Agilent 3000A micro gas chromatograph equipped with multicolumn system, multichannel system and four thermal conductivity detectors (TCD). The liquid products were analyzed based on the boiling point distribution using

Shimadzu GC 2010 Plus with a flame ionization detector (FID) and according to the ASTM D-2887 for standard simulated distillation (SimDist). This boiling point distribution resulted in three liquid fractions defined as gasoline (C5-221°C), light cycle oil-LCO (221–343 °C), and heavy cycle oil-HCO (343+ °C). The results of gas and liquid products analyses were used to calculate conversion and product yields for the three different feeds.

Results and discussion

Effect of temperature on the conversion and total gas yield for the crude oil feeds

The study was carried out using four crude oil samples: a) Arabian Light (AL); b) AL 371-540; c) AL 200+ and d) AL -200+AL540+ Blend. Effect of temperature was studied over 4 temperatures: 600, 625, 650 and 675 °C. Figure 1 shows the effect of temperature on conversion and total gas yield for the four feeds (a-d). It was observed that the AL -200+AL540+ Blend gives the highest conversion and relatively low total gas yield. AL crude oil seems to have higher yield of total gas and high conversion.

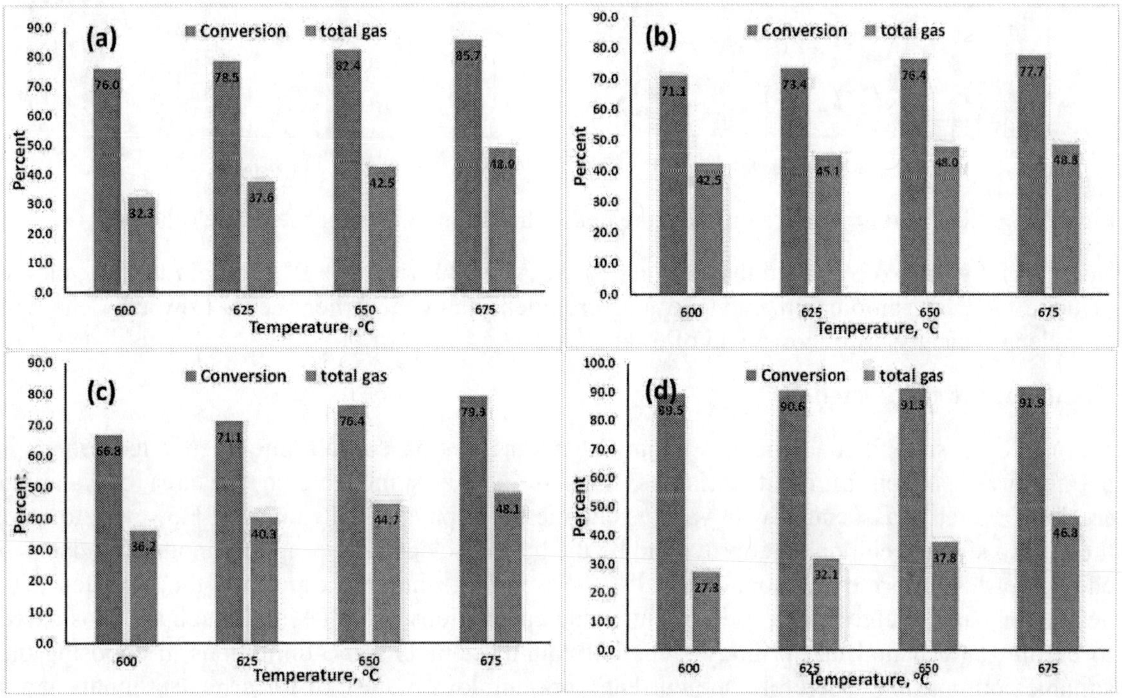

Figure 1. Effect of temperature on conversion and total gas yield for the four (a-d) crude oil feeds

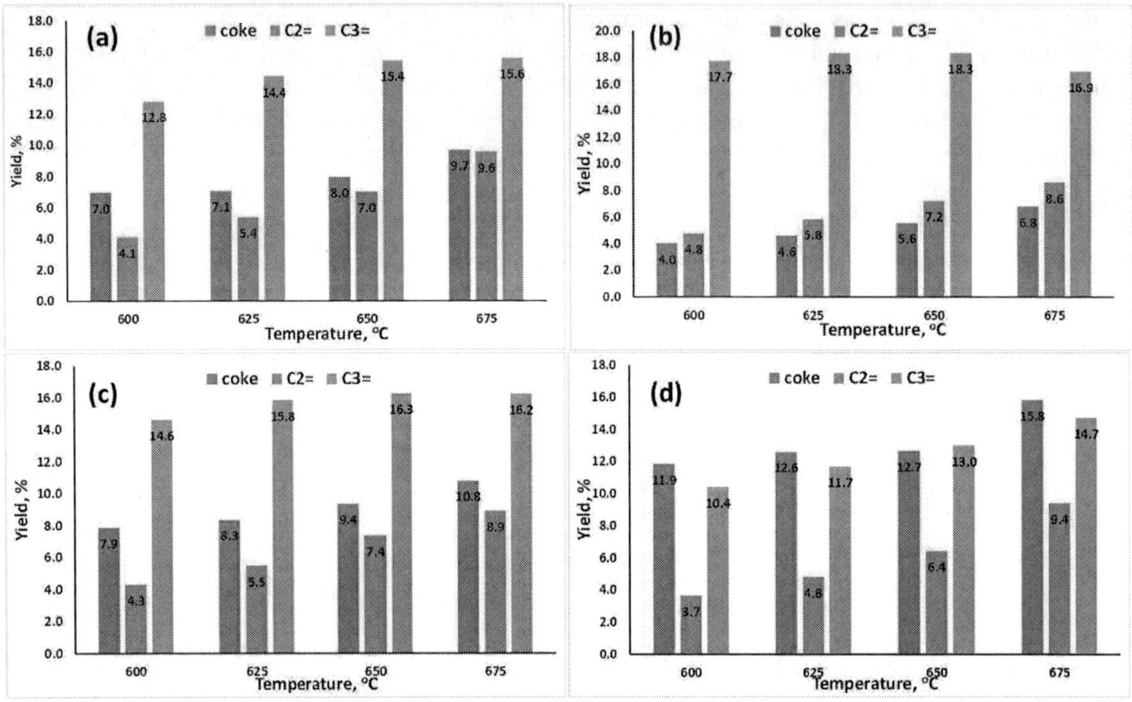

Figure 2. Effect of temperature on the coke and light olefins yield for the crude oil feeds

However, it was also observed that (Figure 2) the AL -200+AL540+ Blend yield largest amount of coke, and low amount of ethylene and propylene than the other feeds. Low ethylene and propylene typically imply low yield of total gas.

Results of the single models

Among the basic and fundamentals of any chemometrics based data-intelligence techniques is the feature selection of input variables. This is equally applicable in the area of catalytic cracking, which is associated with various chemical and physical phenomena. However, to date, there is no single technique that proved to be the best in selecting appropriate input candidates in chemometrics feature extraction. Generally, most of the techniques especially those related with descriptor-based method composed of clustering calculations, which even though are considered to be the fundamental and primary steps but equally depicts some limitations in choosing the suitable composition. Therefore, the current work employs the use of linear matrix input-output based feature extraction technique (see Figure 3).

The current section presents and described the obtained results from both the machine learning based techniques and the Ensemble computational technique for the prediction of C_2H_4 and C_3H_6 concentrations obtained from catalytic conversion. The prediction of the MLs technique was done on MATLAB. Four different grading metrics namely; R^2, PCC, RMSE and MSE were used in evaluating the models.

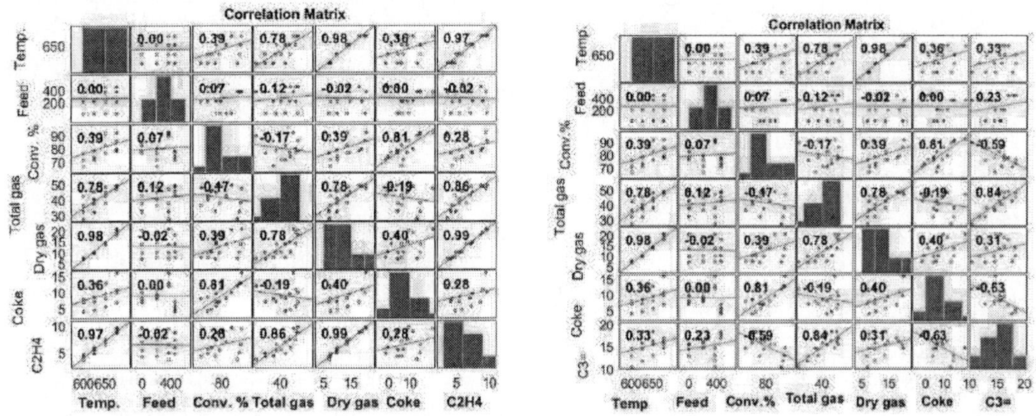

Figure 3. Input-out Feature extraction

The results indicate that the models failed at modeling the C_2H_4 and C_3H_6 concentrations (Figure 4). Hence, there is need for using advance computational techniques such as Ensemble paradigms, hybrid methods and metaheuristic approach to improve the performance of the single models.

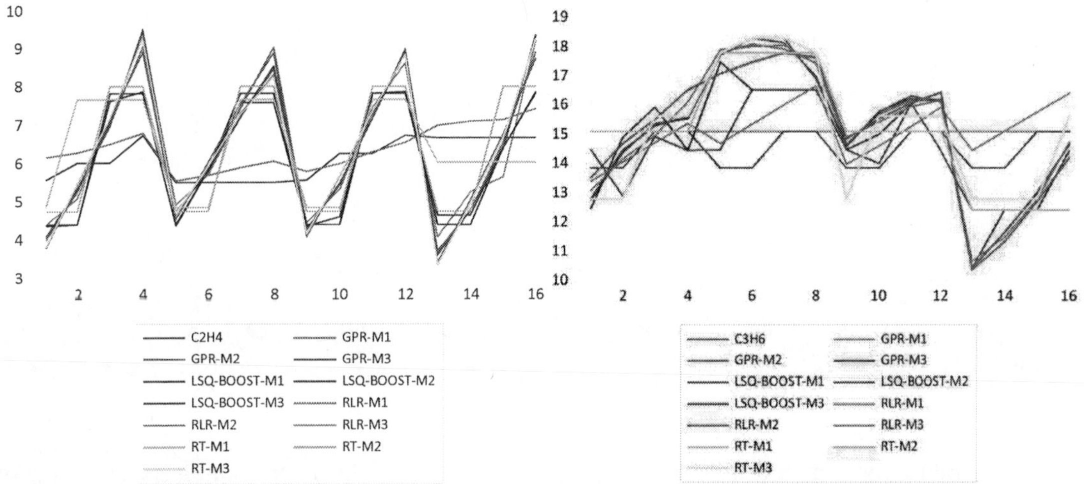

Figure 4. Response plot performance of the single models for C_2H_4 and C_3H_6 concentration modeling

Results of the Ensemble Machine learning paradigms

Owing to the lower accuracy in some instances depicted by the stand-alone models in modeling C_2H_4 and C_3H_6 concentrations, the novel ensemble technique was proposed. The performance accuracy of the ensemble paradigms using ensemble machine learning techniques (GRNN-E, SVR-E, NNE and ANFIS-E) demonstrates the robust application of the novel ensemble technique over the traditional single approaches. This is confirmed by the minimal error obtained from the graphically based error performance plot presented in Figure 5. Hence, the techniques were able to enhance the performance of the models dramatically. Moreover, the fitness comparative performance of the ensemble paradigms can equally be depicted graphically based on time series visualization plots shown in Figure 6.

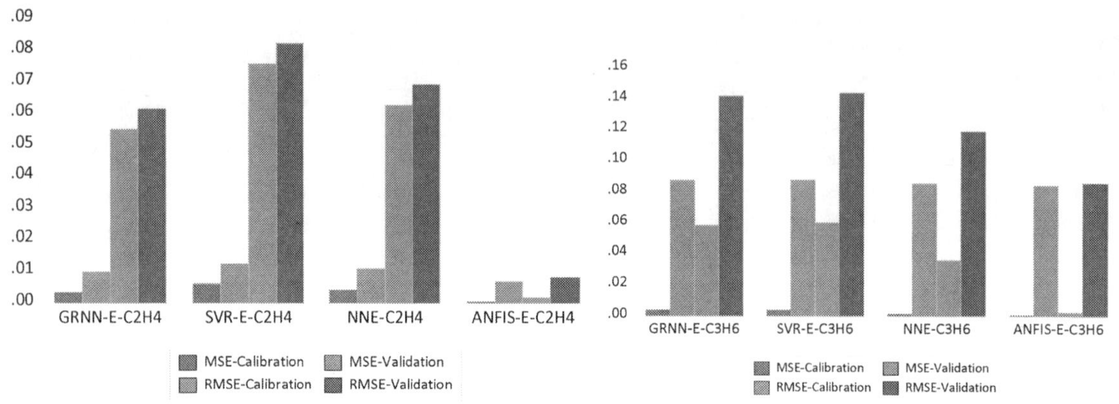

Figure 5. Error performance of the Ensemble paradigms for C_2H_4 and C_3H_6 concentration modeling

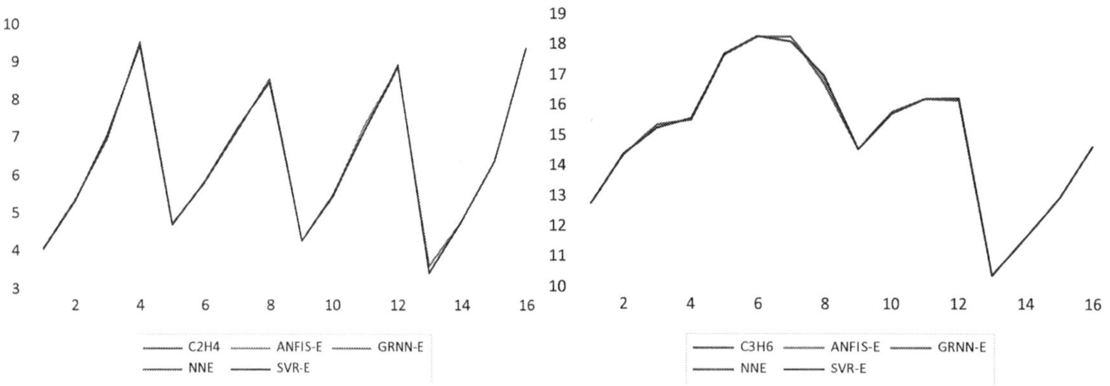

Figure 6. Time series plot of the Ensemble paradigms for C_2H_4 and C_3H_6 concentration modeling

Conclusion

Prediction models for ethylene (C2=) and propylene (C3=) yield during catalytic cracking of various crude oil feeds were developed using temperature, feed, conversion (%), total gas, dry gas and coke as the independent variables. Four stand-alone AI techniques including; Regression tree (RT), least square BOOST (LSQ-BOOST), Gaussian process regression (GPR) and Robust linear regression (RLR) showed weak performance accuracy at certain stages based on the Co-efficient of determination (R^2), Pearson correlation co-efficient (PCC), Mean squared error (MSE) and Root mean squared error (RMSE) performance metrics. An advanced AI technique based on four Ensemble paradigms namely; Generalized Regression Neural Network Ensemble (GRNN-E), Support vector regression ensemble (SVR-E), Feed forward Neural Network Ensemble (NNE) and Adaptive neuro-fuzzy inference system Ensemble improve the performance accuracy of the stand-alone paradigms by up to 58% and 62% in the calibration and validation phases respectively.

References

[1] Nagpal S. Crude Oil Conversion to Chemicals-IHS Markit Process Economics Program. 2021.

[2] Zhou X, Sun Z, Yan H, Feng X, Zhao H, Liu Y, et al. Produce petrochemicals directly from crude oil catalytic cracking, a techno-economic analysis and life cycle society-environment assessment. J Clean Prod 2021;308:127283. https://doi.org/10.1016/J.JCLEPRO.2021.127283.

[3] Usman A, Siddiqui MAB, Hussain A, Aitani A, Al-Khattaf S. Catalytic cracking of crude oil to light olefins and naphtha: Experimental and kinetic modeling. Chemical Engineering Research and Design 2017;120:121–37. https://doi.org/10.1016/J.CHERD.2017.01.027.

[4] Al-Khattaf S, Saeed MR, Aitani A, Klein MT. Catalytic Cracking of Light Crude Oil to Light Olefins and Naphtha over E-Cat and MFI: Microactivity Test versus Advanced Cracking Evaluation and the Effect of High Reaction Temperature. Energy & Fuels 2018;32:6189–99. https://doi.org/10.1021/ACS.ENERGYFUELS.8B00691.

[5] Blay V, Louis B, Miravalles R, Yokoi T, Peccatiello KA, Clough M, et al. Engineering zeolites for catalytic cracking to light olefins. ACS Catal 2017;7:6542–66. https://doi.org/10.1021/ACSCATAL.7B02011/ASSET/IMAGES/LARGE/CS-2017-02011B_0014.JPEG.

[6] Kotrel S, Knözinger H, Gates BC. The Haag–Dessau mechanism of protolytic cracking of alkanes. Microporous and Mesoporous Materials 2000;35–36:11–20. https://doi.org/10.1016/S1387-1811(99)00204-8.

[7] Gbadago DQ, Moon J, Kim M, Hwang S. A unified framework for the mathematical modelling, predictive analysis, and optimization of reaction systems using computational fluid dynamics, deep neural network and genetic algorithm: A case of butadiene synthesis. Chemical Engineering Journal 2021;409:128163. https://doi.org/10.1016/J.CEJ.2020.128163.

[8] Bollas GM, Papadokonstantakis S, Michalopoulos J, Arampatzis G, Lappas AA, Vasalos IA, et al. A Computer-Aided Tool for the Simulation and Optimization of the Combined HDS–FCC Processes. Chemical Engineering Research and Design 2004;82:881–94. https://doi.org/10.1205/0263876041596706.

[9] Bollas GM, Papadokonstadakis S, Michalopoulos J, Arampatzis G, Lappas AA, Vasalos IA, et al. Using hybrid neural networks in scaling up an FCC model from a pilot plant to an industrial unit. Chemical Engineering and Processing: Process Intensification 2003;42:697–713. https://doi.org/10.1016/S0255-2701(02)00206-4.

[10] Kawai E, Sato H, Furuichi K, Takatsuka Toru |, Yoshioka T. Maximizing margins and optimizing operational conditions for residue fluid catalytic cracking with an artificial intelligence hybrid reaction model. J Adv Manuf Process 2022;4:e10118. https://doi.org/10.1002/AMP2.10118.

Disruptive water electrolyzer system for cost-effective hydrogen production: Electrolyte engineering

Kazuhiro Takanabe, Department of Chemical System Engineering, School of Engineering, The University of Tokyo

Abstract

Water electrolysis using renewable power is a key technology for green hydrogen production. Large-scale water electrolyzers are needed to meet energy demand. In addition to cost reduction of renewable electricity, the capital cost of electrolyzer is one of the important factors to determine the cost of hydrogen. Due to the large scale of the electrolyzer, it is necessary to explore innovative concepts for hydrogen production and consider more disruptive technologies than conventional polymer electrolyte membranes (PEM), which becomes very acidic in contact with electrolyte, or alkaline electrolyzers. Resistance to some ions and other impurities is also important, which determines the durability of the electrolyzer. In this presentation, I will introduce innovative nanostructured electrocatalysts and quantitatively describe the application of neutral pH electrolytes, which contain abundant elements in the electrolyzer and allow safe and durable operation. Thus, the concept of "electrolyte engineering" will be introduced, which offers a universal understanding of electrolytic reactions and the potential for large-scale applications.

Introduction

Active utilization of renewable energy is considered essential for building a recycling-oriented society, but the spatiotemporal maldistribution of its power generation has become an issue. As a solution, methods to convert electrical energy into various energy carriers have been proposed. Hydrogen, a typical carrier, can be produced from ubiquitous water, has a high energy density, and is a raw material for various chemical products. However, according to the literature, hydrogen produced by current water electrolysis systems is estimated to cost \$6.3 kg^{-1}, making it less competitively priced than hydrogen from fossil resources [1]. Most of the cost is attributed to the cost of electricity, but the cost of electricity derived from renewable energy sources has been declining in recent years, and further cost reductions will require lowering the capital cost of water electrolysis systems. The electrolytes used in existing water electrolysis systems are strongly basic or acidic, so highly corrosion-resistant materials are used, resulting in electrolysis stack costs (including electrolysis cell components and electrode materials) [2], which account for about half of the system cost. As a measure to reduce such costs, we have focused on the use of neutral pH electrolytes.

Recent studies have shown that when a neutral pH unbuffered aqueous solution is used as an electrolyte, a pH gradient is generated as the reaction proceeds, resulting in a significant decrease in water splitting efficiency. The pH gradient can be minimized when a buffered solution such as phosphate, carbonate or borate is used as an electrolyte [3]. However, mass transfer of buffer ions such as phosphate species contributes to electrolytic performance, and improved performance has been reported by adjusting solution properties to maximize the mass transfer flux [4]. All of these reports are based on room temperature and low reaction rates, and there is a large discrepancy with actual operating conditions.

Therefore, this study focuses on water electrolysis systems that use neutral pH aqueous solutions as electrolytes, and aims to further deepen the fundamental understanding of the systems and their high performance for their practical use. The key to water electrolysis is the development of electrode materials with high activity and the construction of an electrolyte system that minimizes losses due to mass transfer. Furthermore, it is necessary to design a reaction system in which products are not consumed by reverse reactions. We first focus on the physical properties of the electrolyte and report the results of measuring and analyzing the physicochemical properties of the electrolyte under high concentration and high temperature conditions and quantitatively evaluating the mass transfer loss.

Experimental

The physical properties of various aqueous M-phosphate solutions (M = Li, Na, K, Cs) adjusted to pH 7 at 25 °C were measured. Conductivity was evaluated by electrochemical impedance method using a potentiostat (VMP3, Biologic Science Instruments) and viscosity was measured using a viscometer (SVM3001, Anton Paar).

Water electrolysis tests were performed using model electrodes prepared according to previous reports: IrO_x electrodes were prepared using 1 × 1 cm^2 Ti mesh (Fuel Cell Store) as a substrate and electrodeposition solution (0.4 mM $Na_3IrCl_6{\cdot}xH_2O$, 2 mM $H_2C_2O_4$ mixed solution; pH was adjusted to 10 with $Na_2CO_3{\cdot}H_2O$) at a constant current density of 140 μA cm^{-2} for 14 × 10^3 s [5]. Pt electrodes were prepared by applying a constant current density of 140 μA cm^{-2} for 14 × 10^3 s in a mixture of $Na_2CO_3{\cdot}H_2O$ [5]. They were prepared by applying a constant potential of −0.1 V vs. Hg/Hg_2Cl_2 for 15 min in an electrodeposition solution (a mixture of 0.01 M H_2PtCl_6 and 0.023 M $HClO_4$) on a 1 × 1 cm^2 Pt mesh (Nilaco, 99.98%) as substrate. Created by. [6]

Water electrolysis tests were performed by two-electrode and three-electrode methods, with Hg/Hg_2Cl_2 (saturated KCl internal solution) used as the reference electrode in the case of the three-electrode method. Ar (99.9999%) or O_2 (99.9995%) was continuously supplied to the cell before the start of measurement. A 550 mm long glass tube was placed at the top of the cell to provide reflux function. Cyclic voltammetry (CV) and chronopotentiometry (CP) were performed using potentiostats. Cell temperature was controlled by a coolant and hot water circulation system (NCB-1210, Eyela) and an oil bath.

Results and Discussion

In this study, physicochemical properties of M-phosphate buffer solution are first measured at various concentrations and temperatures to quantitatively discuss the mass transfer loss associated with water electrolysis. The mass transfer flux J is generally expressed by the Nernst-Planck equation shown below.

$$J_i(x) = -D_i \frac{\partial c(x)}{\partial x} - \frac{z_i F}{RT} D_i c \frac{\partial f(x)}{\partial x} - cv(x) \quad (1)$$

D is the diffusion coefficient, *c* is the solute concentration, *z* is the ionic valence, *F* is Faraday's constant, *R* is the gas constant, *T* is the temperature, and *v* is the velocity of the fluid. The diffusion coefficient *D* in equation (1) is expressed in dilute solution by the following Stokes-Einstein equation based on the rigid sphere model.

$$D = \frac{kT}{3\pi d\mu} \quad (2)$$

where *k* is Boltzmann's constant, *d* is Stokes diameter, and μ is solution viscosity. From equation (2) and the mobility equation, conductivity can be expressed as

$$\sigma = \frac{z^2 ecF}{3\pi d\mu} \quad (3)$$

where *e* is the elementary charge. We first measured solution viscosity and conductivity at various concentrations and temperatures to determine the applicability limits of this model. The viscosity measurements of Na-, K-, and Cs-phosphate solutions at concentrations of 2.0 mol kg^{-1} are shown in Figure 1(a). The results show that regardless of the cation species, the concentration of concentrated phosphate buffer solution decreases with increasing temperature. The Na- and Cs-phosphate solutions had a solution viscosity of 1.1 mPa s at 100 °C, whereas the K-phosphate solution had a lower value of 0.8 mPa s. Next, conductivity was calculated from the measured solution viscosity using equation (2) and compared to the measured conductivity. Figure 1(b) shows the results of the comparison between the calculated and measured values. It is clear that the measured and calculated conductivities are in quantitative agreement at concentrations ranging from 0.5 to 4.1 mol kg^{-1}, indicating that the description based on the rigid sphere model is applicable even to such concentrated buffer solutions. Therefore, it is suggested that the K-phosphate solution, which has the lowest viscosity among the M-phosphate solutions examined in this study, may have the highest diffusion coefficient value and therefore the lowest mass transfer loss.

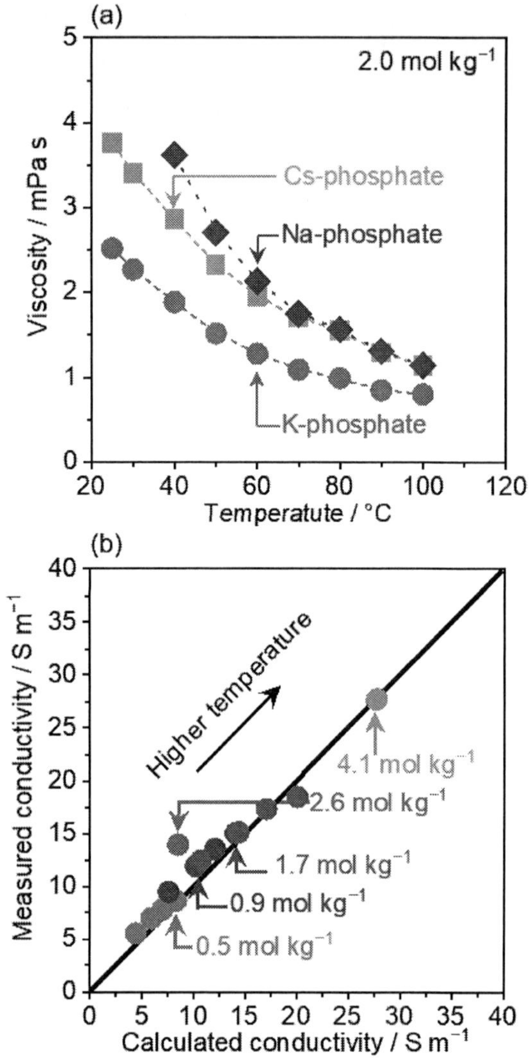

Figure 1: Results of physical properties measurements. (a) Results of solution viscosity measurements. Solution: 2.0 mol kg^{-1} Na-, K-, Cs-phosphate solution. (b) Comparison of measured and calculated conductivity. Measurements were taken with 0.5, 0.9, and 1.7 mol kg^{-1} at 25, 40, 50, and 60 °C, 2.6 mol kg^{-1} at 25 °C, and 4.1 mol kg^{-1} at 100 °C in K-phosphate solutions. Calculated values are calculated using equation (2).

The physicochemical data obtained were then used to evaluate the mass transfer loss related to water electrolysis. Figure 2 shows the results of the calculation of iR loss due to swimming and concentration overvoltage due to diffusion in the case of K-phosphate solution. For the calculations, a current density of 100 mA cm^{-2} and a distance of 0.5 mm between electrodes were assumed. In the iR loss analysis shown in Figure 2(a), the iR loss for saturated K-phosphate solution decreased with increasing temperature and reached about 90 mV at 100 °C, which is larger than the value of 15 mV. In an actual system, a separation membrane is installed in the electrolyte, and when its electrical resistance is considered, the iR loss in 7.0 mol kg^{-1} KOH solution at 80 °C increases to 135 mV. On the other hand, in a system using a concentrated buffer solution, it has been proposed that the crossover of the product gas is suppressed at 25 °C and the separation membrane may be omitted. [7] Therefore, if crossover is similarly suppressed under elevated temperature conditions, it is suggested that a system using a K-phosphate solution could achieve lower iR loss than existing alkaline electrolysis. Figure 2(b) shows the results of the concentration overvoltage analysis, which indicates that, as with the iR loss, the concentration overvoltage decreases with increasing temperature in a K-phosphate solution, and can be as low as 20 mV at 100 °C. However, the value of the concentration overvoltage is not statistically significant compared to that of the iR loss. However, the value of concentration overvoltage is about one order of magnitude smaller than that of iR loss, indicating that diffusion flux is the dominant mass transfer in water electrolysis using high temperature and concentrated buffer solutions as electrolytes, and that its increase is a more significant issue in this reaction.

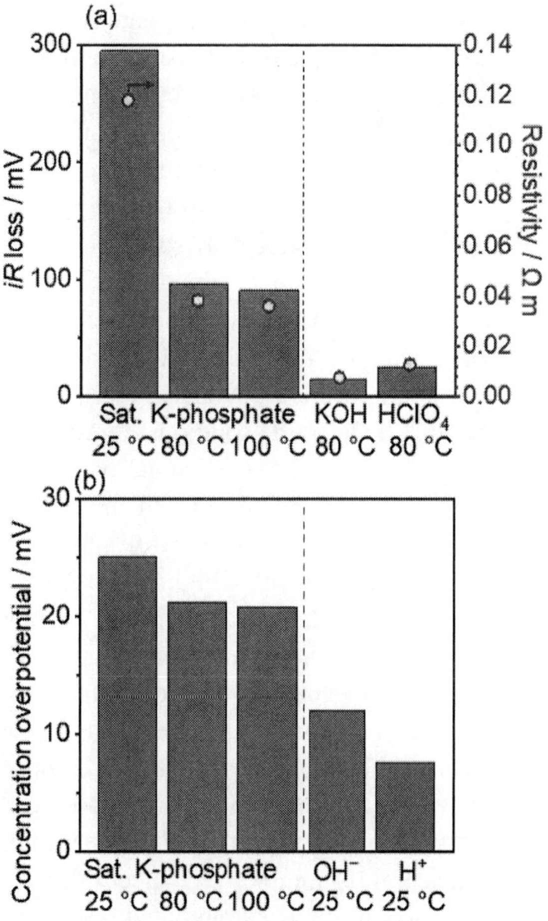

Figure 2. Voltage loss due to mass transfer when 100 mA cm^{-2} is applied. (a) iR loss analysis results. Resistivity was calculated by the impedance method using a two-electrode method with platinum electrodes; KOH is a reference value [8]. (b) Concentration overvoltage analysis results. Electrode surface area: 1.0 cm^2, diffusion layer: 0.5 mm. electrolyte concentration: 2.6 mol kg^{-1} at 25 °C, 3.5 mol kg^{-1} at 80 °C, 4.1 mol kg^{-1} at 100 °C for K-phosphate (pH 7.0). $HClO_4$ is 7.0 mol kg^{-1} and KOH is 7.1 mol kg^{-1}. OH^- is 1.0 mol kg^{-1} and H^+ is 1.0 mol kg^{-1}. OH^- and H^+ are reference values [8].

Based on these results, water electrolysis tests were conducted using saturated K-phosphate buffer solution as the electrolyte. Pt/Pt mesh electrode for the cathode and IrO_x/Ti mesh electrode for the anode were used as model electrodes for the tests. Figure 3(a) shows the CP measurement results at 10 mA cm^{-2} and 80 °C. The electrolyte was saturated K In addition to saturated K-phosphate solution, 7.0 mol kg^{-1} $HClO_4$ and KOH solutions were used as electrolytes for comparison. Immediately after the start of the test, the iR correction voltages in the $HClO_4$ and KOH solutions were 1.24 V and 1.30 V, respectively, which were lower than the 1.44 V of the K-phosphate solution, but both increased with time, reaching >2.0 V and >1.6 V, respectively. On the other hand, the iR-corrected voltage in the K-phosphate solution was 1.48 V after 6 h, indicating that it was more stable than the $HClO_4$ and KOH solutions. Figure 3(b) shows the electrolysis efficiency at 100 °C in saturated K-phosphate solution and the performance of existing electrolysis systems [9, 10] comparison results. In this figure, the dotted line represents the extrapolation of the practical electrolytic efficiency in K-phosphate solution with oxygen evolution reaction (OER), hydrogen evolution reaction (HER) overvoltage, plus iR loss. The iR loss was calculated by multiplying the measured conductivity of the saturated K-phosphate solution by the cell constant for a distance of 0.5 mm between electrodes. Comparison of this extrapolated line with the efficiency of existing electrolyzers indicates that the electrolytic performance of the model electrode in the high temperature and high concentration of K-phosphate studied in this study is comparable to that of existing electrolyzers.

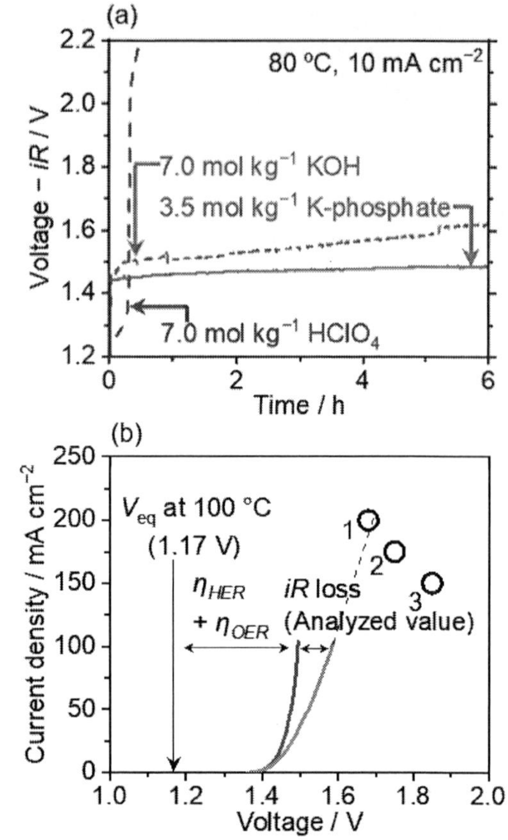

Figure 3. Water electrolysis performance in concentrated buffer solution. (a) CP measurements at 10 mA cm^{-2}. Electrolytes: 7.0 mol kg^{-1} KOH, 7.0 mol kg^{-1} $HClO_4$, 3.5 mol kg^{-1} K-phosphate, 80 °C. (b) CV measurement results. Electrolyte: 4.1 mol kg^{-1} K-phosphate, scan rate: 10 mV s^{-1}, 100 °C. 1-3 shows current electrolyzer performance. [9, 10] All measurements were performed under Ar atmosphere, with IrO_x/Ti mesh anode and Pt/Pt mesh cathode.

Conclusions

In this study, physicochemical properties of concentrated phosphate solution adjusted to pH 7 as an electrolyte for water electrolysis were measured at various temperatures and applied to the water electrolysis reaction. First, the agreement between the measured and theoretically calculated conductivity values indicates that the existing model for dilute solutions is applicable to the concentrated solution used in this study. In addition, the calculation results of mass transfer flux during water electrolysis showed that the electrolysis system using high-concentration K-phosphate solution at high temperature exhibits a mass transfer flux equivalent to that of existing strong acid and strong alkaline electrolyzers. Furthermore, in a demonstration test of water electrolysis at high temperature using model electrodes, it was found that the electrocatalytic efficiency is comparable to that using strong acid and strong alkali electrolyte solutions, and that higher stability is achieved. These results suggest that neutral pH concentrated buffer solution is promising as an electrolyte for water electrolysis [11]. Ongoing study includes evaluation of the crossover of the produced gases in the concentrated buffer solution under high temperature conditions [12]. We are also working on developing a highly efficient OER catalyst comprising of non-noble metal to improve the performance of water electrolysis in concentrated buffer solution.

References

[1] S.E. Hosseini, et al., *Renewable Sustainable Energy Rev.* **2016**, *57*, 850.

[2] L. Bartuccioli, et al., "*Development of water electrolysis in the European Union*", Fuel Cells and Hydrogen Joint Undertaking, **2014**.

[3] I. Katsounaros, et al., *Electrochem. Commun.* **2011**, *13*, 634-637.

[4] T. Shinagawa, et al., *ChemSusChem* **2017**, *10*, 1318-1336.

[5] M. A. Petit, et al., *J. Electroanal. Chem.* **1998**, *444*, 247-252.

[6] A. M. Feltham, et al., *Chem. Rev.* **1971**, *71*, 177-193.

[7] T. Shinagawa, et al., *J. Phys. Chem. C* **2016**, *120*, 1785−1794.

[8] W.M. Haynes, et al., *Handbook of Chemistry and Physics, 92nd ed.*, CRC Press, Boca Raton, FL, **2011**.

[9] R. E. White, et al., *Modern aspects of electrochemistry No.15*, Prenum Press, New York & London, **1983**.

[10] K. Zeng, et al., *Prog. Energy Cmbus. Sci.* **2010**, *36*, 307-326.

[11] T. Nishimoto, et al., *ChemSusChem* **2021**, *14*, 1554-1564.

[12] T. Naito, et al., *ChemSusChem* **2022**, *15*, e202102294.

Development of low temperature steam reforming catalyst for hydrogen production for fuel cells

Shakeel Ahmed[1], Uwais Baduruthamal[1], Sai P. Katikaneni[2], and Aadesh Harale[2]

[1] Center for Refining and Petrochemicals, Research Institute, King Fahd University of Petroleum & Minerals, Dhahran-31261, Saudi Arabia.

[2] Research & Development Center, Saudi Arabian Oil Company (Saudi Aramco), Dhahran, Saudi Arabia.

A series of selected lanthanides (La, Ce, Pr, Sm, Eu, and Gd) modified Ni/Al_2O_3 catalysts was prepared by impregnation technique for low temperature methane steam reforming to produce hydrogen for fuel cells. According to the thermodynamic equilibrium analysis, it is possible to operate methane steam reforming at low temperatures. Structured low temperature steam reforming catalyst in combination with hydrogen selective membrane presents great potential of intensifying the classical industrial hydrogen production process via natural gas. A highly active catalyst is required for the production of hydrogen at lower temperatures. The catalysts prepared in this study were characterized by X-ray diffraction, Brunauer−Emmett−Teller (BET), transmission electron microscopy (TEM), scanning electron microscopy, energy dispersive X-ray, and temperature-programmed reduction (TPR) characterization techniques. The performance evaluation of the prepared catalysts was conducted in a fixed bed micro-reactor at steam to carbon ratio of 3.0 and at a pressure of 2.1 bar over a range of temperature 500 to 600 C. The results indicated that Pr modified Ni/Al_2O_3 catalyst performed better in terms of conversion of methane to other catalysts in the series. Further investigations were carried out by varying the amount of Pr over a range of 1 to 7 wt% and found that 3% Pr modified 12% Ni/Al_2O_3 catalyst showed better performance for the SMR reaction. Pr− Ni catalyst has the higher BET surface area among other synthesized lanthanide-promoted 12%Ni/γ-Al_2O_3 catalysts. TEM analysis indicated uniform metal dispersion of promoted catalysts. Moreover, TPR analysis indicated an optimum interaction between the metal and the support in the Pr− Ni catalyst.

Thermo-catalytic production of hydrogen from H_2S: Challenges and Opportunities

Hassan Aljama, Zainab Al-Aithan and Ali Almofleh

R&DC, Saudi Aramco

H_2S is a toxic and corrosive substance that can pose great risks to human health, even upon exposure at low concentrations. It is mostly produced as a by-product of industrial processes, especially in the oil and gas industry. Worldwide, H_2S production exceeds 10 million tons per year. Strict environmental regulations require >99% sulfur recovery. To meet these stringent requirements, the current industrial practice is to use the Claus process to convert H_2S to elemental sulfur and water. Ideally, hydrogen in H_2S is recovered as H_2, given that hydrogen is a valuable commodity with many important applications, especially in the energy and chemical sectors, and the expected growth in demand for hydrogen over the coming years. Several attempts were made in the literature to achieve this objective dating back to 1939. Despite much effort in the area, especially in thermal catalysis, a commercial application remains distant. A fundamental factor is the lack of an effective catalyst. In this work, we employ density functional theory (DFT) to illustrate the main limitations in existing catalysts. We use pure metals to explain this by studying the full elementary steps in H_2S decomposition. We find that many catalysts, though capable of decomposing H_2S, are limited due to sulfur poisoning. We conclude by outlining the ideal properties of a catalyst for this process.

Titania supported bimetallic catalyst for hydrogen production from methane

Wasim Ullah Khan[1], Hassan Alasiri[1,2], Mohammad Mozahar Hossain[1,2]

[1]Interdiscipilinary Research Centre for Refining & Advanced Chemicals, Research Institute, King Fahd University of Petroleum & Minerals, Dhahran 31261, Saudi Arabia

[2]Department of Chemical Engineering, King Fahd University of Petroleum and Minerals, Dhahran, 31261, Saudi Arabia

The synthesis of hydrogen, a clean energy source, has recently drawn the attention of scientists. Hydrogen, as an environmentally friendly energy source, finds its utilization in various applications including spaceships, fuel cells, and converting heavy petroleum fractions to produce lighter ones. Conventional hydrogen production processes include steam methane reforming, dry or (carbon dioxide) reforming of methane, and/or partial oxidation of methane. However, these hydrogen production processes have drawbacks such as high carbon footprint, carbon deposition, and/or challenging temperature control, etc.,. Catalytic decomposition and/or cracking of methane offers an alternative process to the above-mentioned processes with the advantage of zero-emission hydrogen production along with valuable byproducts in the form of carbon nanomaterials. Hence, this work highlights the role of nickel and cobalt-based bimetallic catalysts supported on titanium dioxide (TiO_2) during the catalytic decomposition of methane. Bimetallic catalysts with various bimetal loading ranging from 20 to 60 wt% are tested at 700°C at atmospheric pressure. Bimetallic catalyst with 25 wt% each of nickel (Ni) and cobalt (Co) has demonstrated the highest methane conversion and hence remains the optimum loaded catalyst. The catalysts are also characterized to explore the impact of metal-support interaction, specific surface area, and crystalline structure on catalytic performance.

The activity result with respect to the variation in total metal content (with Ni/Co ratio of 1) from 20 to 60 wt% reveals that metal loading plays a role in influencing methane conversion. For instance, the bimetallic catalyst with 10 wt% each of Ni and Co (dented as 10NCT where N, C, and T represent Ni, Co, and TiO_2, respectively) shows an initial methane conversion of 73% which gradually drops to approximately 70% after 100 min time-on-stream (TOS) and quickly deactivates afterward reaching methane conversion of ~43% after 5 h TOS. A further increase in bimetal content to 20 wt% (20NCT) exhibits a relatively better (stable) methane conversion performance with an initial conversion of ~72% which eventually approaches 81.5% after 5 h TOS. An optimum loading, corresponding to 25NCT, with approximately similar initial methane conversion as of 20NCT demonstrates a final conversion of just over 82% at the end of 5 h of reaction. A further increase in bimetal loading (30NCT) shows nearly 81% final methane conversion. It is interesting to note that lower-loading catalysts (10NCT and 15NCT) deactivated over time while higher-loading catalysts (20NCT, 25NCT, and 30NCT) demonstrated an increase in methane conversion with TOS. This indicates that higher-loading catalysts exhibit an induction period due to the presence of unreduced metal oxides that are getting activated under a reductive environment. Moreover, the loss of active sites due to sintering and coverage of active metals with graphitic carbon are major factors behind catalyst deactivation. Based on catalytic activity results, it can be inferred that 25NCT catalyst outperforms the rest of the catalysts and this catalytic performance can be assigned to relatively higher specific surface area (115 m^2/g), suitable metal-support interaction and extent of reducibility.

Ammonia to hydrogen: Cracking technology

Sai P. Katikaneni, Mohammad A. Rakib, Kunho Lee, Stephen N. Paglieri, Bandar H. Solami, Aqil Jamal

Aramco Research & Development Center, Dhahran, Saudi Arabia

Hydrogen can play a vital role in decarbonizing our current energy system, although several challenges related to its production, distribution, and storage must be overcome. Ammonia is one of the most important practical hydrogen carriers over long distances due to its high energy density, relatively low cost per unit of energy stored, well-defined regulations, solid safety record, and well-established production technology and distribution networks.

Aramco is developing an efficient catalyst that will crack ammonia to make hydrogen in an efficient heat optimized reactor. Two types of ammonia cracking catalysts, a K-promoted Ru-based catalyst and a Ba-promoted Co-Ce ammonia cracking catalyst, have been developed, with both exhibiting high activity and complete ammonia conversion at lower temperatures than conventional catalysts (e.g. based on nickel). Catalyst manufacturing has been scaled up to kilogram-scale batches, and its performance has been validated at the pilot scale.

Ammonia cracking is an endothermic, equilibrium constrained process that is favored at higher temperatures and lower pressures. Furthermore, a large heat source is required to maintain optimal reaction temperature. Separation of any reaction product improves reaction conversion. Utilizing H_2-selective Pd-based membranes in conjunction with a fixed bed of catalyst-filled tubes increased the hydrogen yield , along with effective validation of the scaled-up catalyst in the membrane reactor.

The team is scaling up ammonia cracking in a heat-optimized multi-tubular reactor, thereby demonstrating this technology as a key enabler for exporting hydrocarbon-based hydrogen for emerging transportation, power generation and steel industry applications to meet net zero emission targets.

Redox dehydrogenation of propane utilizing lattice S^{2-} in metal sulfide catalyst

Ryo Watanabe, **Verma Priyanka**, **Choji Fukuhara**
Department of applied Chemistry and Biochemical Engineering, Graduate school of Engineering, Shizuoka university, 3-5-1 Johoku, Naka-ku, Hamamatsu, Shizuoka 432-8561, Japan

Abstract

Dehydrogenation of alkane produces olefine with high selectivity using metal sulfide catalyst. However, there is a crucial problem regarding the loss of a sulfide ion (S^{2-}) from the metal sulfide by H_2 reduction, which leads to decreased dehydrogenation activity. Therefore, suppression of lattice S^{2-} release from metal sulfide is necessary to maintain high dehydrogenation activity. H_2S is known to decompose to S^{2-} on various metal catalysts. We expected that this would improve stability of the catalyst through continuous regeneration of S^{2-} during the reaction. Based on this assumption, the effect of co-supplying H_2S in propane was investigated on transition metal catalysts. The Fe sulfide catalyst displayed high activity and selectivity for the production of propylene; its overall performance was superior to that of the industrial Cr-based catalyst. In addition, Fe sulfide catalyst showed high durability performance for propane dehydrogenation with co-feeding H_2S for 50 h reaction.

Keywords: Redox, Hydrogen sulfide, Metal sulfide, Fe

1. Introduction

Dehydrogenation of propane ($C_3H_8 \rightarrow C_3H_6 + H_2$) has attracted attention because propane, an economical feedstock, can be converted into valuable propylene.[1,2] This reaction needs high temperature and low pressure due to highly endothermic reaction. This reaction is a desirable process because high propylene selectivity can be obtained using Cr- and Pt-based catalysts.[3-5] Coke deposition on these catalysts during the dehydrogenation of propane results in a loss of catalyst performance within a short reaction time.[6] Such short reaction requires a regeneration of the catalysts by coke combustion using diluted air. Such regeneration process for coke removal causes irreversible deactivation of the catalyst due to phase transformation and sintering. Therefore, there is a strong need to develop novel catalysts with stable performance.

Recently, we found that a transition metal (Fe, Co, Ni) catalyst loaded onto aluminum oxide (γ-Al_2O_3) with a sulfate ion (SO_4^{2-}) was selective and relatively stable for dehydrogenation of propane.[7] The loaded sulfate species (SO_4^{2-}) were reduced to sulfide ions (S^{2-}) in the reaction atmosphere and then functioned as the active site for

PDH. However, there is a crucial problem regarding the loss of a sulfide ion (S^{2-}) from the metal sulfide via a reaction with H_2, which leads to decreased dehydrogenation activity.[8] Therefore, the suppression of the release of S species from the metal sulfide is necessary to maintain high dehydrogenation activity. Thus, the objective of this paper is to develop a stable dehydrogenation system by using metal-sulfide catalyst and hydrogen sulfide (H_2S) as co-feeding gas. We expected that this would improve the stability of the catalytic performance of the catalyst through continuous regeneration of S^{2-} during the reaction. Specifically, the effect of the addition of a co-supplying H_2S with propane on the PDH performance of transition metal catalysts (Fe, Co, Ni, Mn, and Cu) was investigated. Then, the performance of the optimized catalyst was compared to that of the Cr-based catalyst. In addition, the durability performance test was performed.

2. Experimental

2.1. Catalyst preparation

The SiO_2 (JRC-SIO-4) support was supplied by the Catalysis Society of Japan. The transition metal oxide (Fe, Co, Ni, Mn, Cu) was loaded using an impregnation method: The corresponding metal nitrate (i.e., $Fe(NO_3)_3$, $Co(NO_3)_2$, $Ni(NO_3)_2$, $Mn(NO_3)_2$, $Cu(NO_3)_2$) and ammonium sulfate ($(NH_4)_2SO_4$) were impregnated on the support and then evaporated until dry and subsequent calcination of the catalyst. For comparison, a $CrO_x/\gamma\text{-}Al_2O_3$ catalyst was prepared using a conventional impregnation method with $Cr(NO_3)_3 \cdot 9H_2O$ (Kanto Kagaku Inc., Japan) and $\gamma\text{-}Al_2O_3$ followed by sequential calcination at 650°C in air. The loadings of Cr components were 10, 15, and 20 wt%, respectively.

2.2. Evaluation of catalyst performance

The PDH performances of the prepared catalysts, including the catalytic activity, selectivity for the product, and stability, were examined in a conventional fixed-bed reactor. The reaction was performed at 600°C under atmospheric pressure. Propane was supplied to the catalyst bed with co-feeding H_2S, as balanced by helium. The molar ratio of H_2S to propane ranged from 0.2 to 2.6. The propane feed rate was 2.5 $mL \cdot min^{-1}$ (SATP), and the catalyst weight was 0.25 g. The gaseous reactants and products in the effluent gases (i.e., C_3H_8, CO, CH_4, C_2H_4, C_2H_6, and C_3H_6) were collected by micro-syringe and subsequently injected into the off-line thermal conductivity detection gas chromatograph and flame ionization detection gas chromatograph (GC-8A; Shimadzu Inc., Japan).

3. Results and discussion

3.1. Effect of transition metal species[9]

The loss of lattice S^{2-} from the metal sulfide via reaction with H_2 and/or propane leading to decreased dehydrogenation activity is a critical issue. Suppression of the loss of S^{2-} from the metal sulfide could lead to a high dehydrogenation performance. Therefore, the dehydrogenation of propane in the presence of H_2S was investigated. The catalytic performance was investigated for propane dehydrogenation with co-feeding of H_2S over SiO_2-supported transition metal (Fe, Co, Ni, Mn, Cu) catalysts (described henceforth as Metal/SiO_2). Figure 1(a) shows the conversion of propane over the Metal/SiO_2 catalysts. The Fe/SiO_2, Co/SiO_2, and Ni/SiO_2 catalysts showed high propane conversion at the initial stage of the reaction. In addition, the Fe/SiO_2 catalyst displayed better stability during propane dehydrogenation with co-feeding of H_2S than the other transition metal catalysts. The Mn/SiO_2 and Cu/SiO_2 catalysts showed low propane conversions. The product selectivity at 200 min is shown in Figure 1(b). From the figure, propylene, methane, and ethylene were confirmed as the products formed over all the catalysts. CH_4 and C_2H_4 by-products were produced by decomposition of propane ($C_3H_8 \rightarrow CH_4 + C_2H_4$) and/or hydrogenolysis of propylene ($C_3H_6 + H_2 \rightarrow CH_4 + C_2H_4$). A comparison of the propylene selectivity over each catalyst revealed that the Fe/SiO_2 catalyst displayed the highest selectivity of *ca.* 93%; this high selectivity almost remained constant throughout the dehydrogenation.

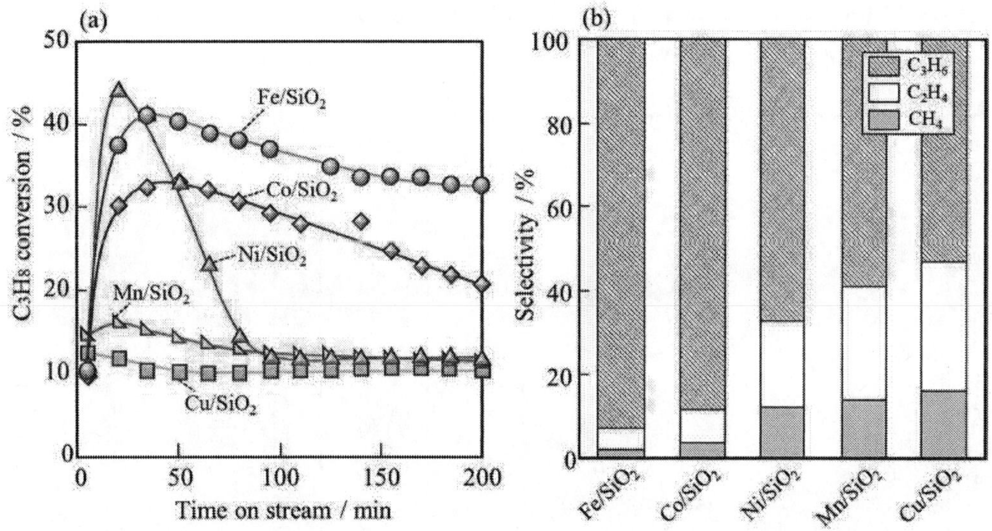

Fig. 1 (a) Propane conversion and (b) product selectivity over Metal/SiO_2 catalysts (Metal: Fe, Co, Ni, Mn, and Cu) with co-feeding H_2S.[9]

3.2. Effect of type of support[10]

Effect of the type of support (γ-Al_2O_3, SiO_2, ZrO_2, MgO, and CeO_2) on the performance of the Fe-based catalyst for propane dehydrogenation with co-feeding of H_2S was investigated. Figure 2(a) shows the conversion of propane over the Fe-based catalysts with various supports. The value in parenthesis in Fig. 2(a) shows the amount of coke deposited after propane dehydrogenation with co-feeding H_2S. The supported catalyst is described as Fe/support henceforth. From Figure 2(a), it is clear that the Fe/γ-Al_2O_3 and Fe/SiO_2 catalysts showed high propane conversion after 35 min of reaction. In addition, the Fe/SiO_2 catalyst displayed better stability for propane dehydrogenation with co-feeding H_2S than the Fe/γ-Al_2O_3 catalyst. The Fe/ZrO_2 and Fe/CeO_2 catalysts showed catalytic activity to a certain degree; however, these catalysts were deactivated significantly. In contrast, the Fe/MgO catalyst exhibited no catalytic activity for propane dehydrogenation with co-feeding H_2S. Therefore, the stability of these catalysts was largely dependent on the type of support used. Figure 2(b) shows the product selectivity after 200 min of reaction. The Fe/SiO_2 catalyst showed the highest propylene selectivity. The propylene selectivities of the other catalysts, that is, the Fe/γ-Al_2O_3, Fe/ZrO_2, Fe/MgO, and Fe/CeO_2 catalysts were 82.7%, 74.5%, 73.0%, and 67.3%, respectively. The difference in selectivity could be attributed to the effect of the properties of the support on the dehydrogenation reaction. The γ-Al_2O_3, ZrO_2, and CeO_2 catalyst supports contain electron-deficient cations that act as Lewis acid sites[31)]. The propylene molecule contains a π-bond which makes it more basic than propane. Therefore, propylene exhibits stronger interactions with the electron-deficient Lewis centers on the oxide surfaces than with propane molecules. This basicity leads to the cleavage of the allylic C–C bonds in propylene, which may decrease the propylene selectivity.

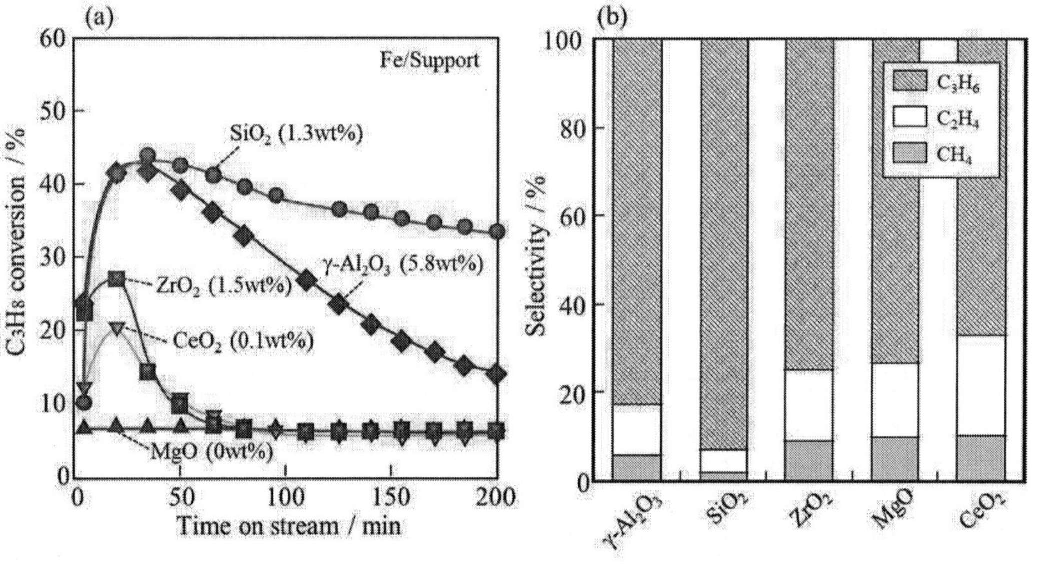

Fig. 2 (a) Propane conversion and (b) product selectivity over Fe catalysts on various supports (SiO_2, γ-Al_2O_3, ZrO_2, MgO, and CeO_2).[10]

3.3. Practicability of Fe-based catalyst

The performance of the Fe/SiO_2 catalyst was compared to that of the CrO_x/γ-Al_2O_3 catalyst, which is known to be a highly active catalyst for propane dehydrogenation. Figure 3 shows the propylene yields as a function of the reaction time. The CrO_x loadings of 10, 15, and 20 wt% were investigated. The Fe/SiO_2 catalyst exhibited less activity than the $CrO_x/\gamma\text{-}Al_2O_3$ catalyst. However, the stability of the Fe/SiO_2 catalyst was superior to that of the $CrO_x/\gamma\text{-}Al_2O_3$ catalyst.

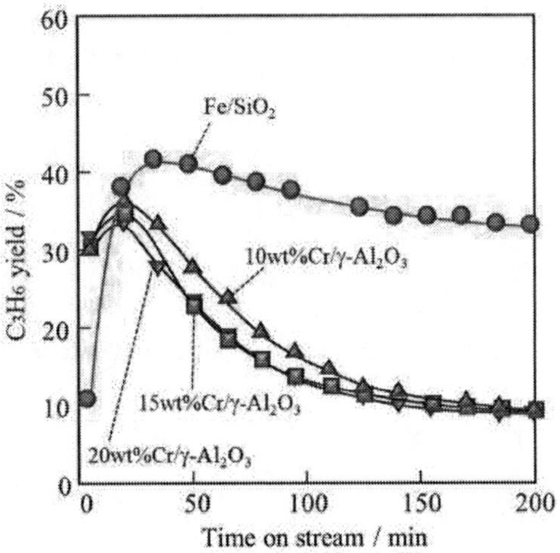

Fig.3 Effect of the addition of H_2S on the C_3H_6 yield of the Fe/SiO_2 and $Cr/\gamma\text{-}Al_2O_3$ catalysts. The reaction conditions were as follows: reaction temperature was 600°C, catalyst weight was 250 mg, and the ratio of $H_2S/C_3H_8/He$ flow rates was 3.5:2.5:20 $mL\cdot min^{-1}$.[9]

Figure 4 shows the durability of the Fe/SiO_2 catalyst during a reaction period of 50 h using H_2S/C_3 values of 0.2, 0.4, and 1.4. The initial propylene yield was high at a H_2S/C_3 ratio of 1.4. However, the catalyst underwent deactivation as the reaction proceeded. After 35 min, the conversion was 49.5% with 94.4% propylene selectivity; the activity decreased, exhibiting a conversion of 15.0% with 78.2% propylene selectivity after 50 h. This result indicates that the catalyst degraded significantly during the reaction. At H_2S/C_3 values of 0.2 and 0.4, the initial conversion was lower than that at the H_2S/C_3 value of 1.4. However, only a slight decrease in the conversion and selectivity was observed. Additionally, the catalyst underwent marginal deactivation over the 50-h reaction period at H_2S/C_3 values of 0.2 and 0.4, that is, a relatively high conversion and up to 90% selectivity were observed at a H_2S/C_3 ratio of 0.4.

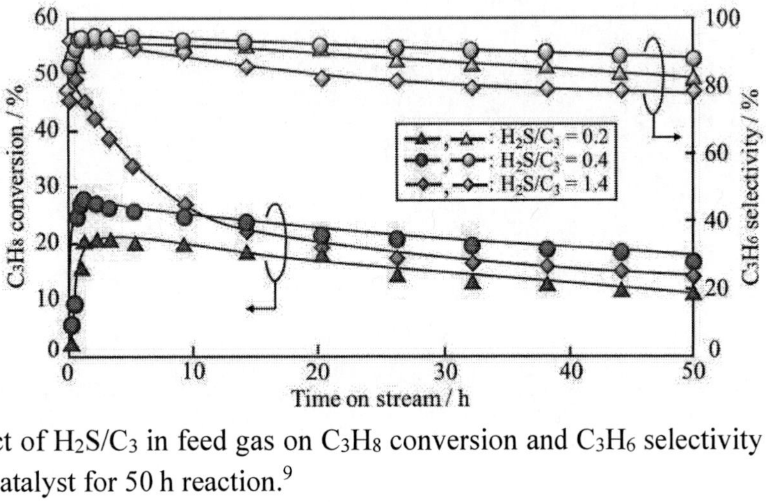

Fig. Effect of H_2S/C_3 in feed gas on C_3H_8 conversion and C_3H_6 selectivity over the Fe/SiO_2 catalyst for 50 h reaction.[9]

4. Conclusions

The high activity was attributed to the S^{2-} ions formed by reduction during the reaction. Metal sulfides were effective catalysts for the dehydrogenation of propane; however, the loss of S^{2-} ions from the catalyst surface under the reaction atmosphere employed degraded the catalytic performance. During the dehydrogenation reaction in the presence of H_2S, the S^{2-} ions were continuously restored on the catalyst surface, and high stability was achieved.

References

1) A. Corma, F.V. Melo, L. Sauvanaud, F. Ortega, *Catal. Today*, 107 (2005) 699-706.
2) J.S. Plotkin, *Catal. Today*, 106 (2005) 10-14.
3) F.E. Frey, W.F. Huppke, *Ind. Eng. Chem.*, 25 (1) (1933) 54-59.
4) S. De Rossi, G. Ferraris, S. Fremiotti, E. Garrone, G. Ghiotti, M.C. Campa, V. Indovina, J. Catal., 148 (1) (1994) 36-46.
5) K. Xia, W.-Z. Lang, P.-P. Li, X. Yan, Y.-J. Guo, *J. Catal.*, 338 (2016) 104-114.
6) I. Miracca, L. Piovesan, *Catal. Today*, 52 (2–3) (1999) 259-269.
7) R. Watanabe, N. Hirata, C. Fukuhara, *J. Jpn. Petrol. Inst.*, 60(5) (2017) 223-231.
8) G. Wang, C. Li, H. Shan, *ACS Catal.*, 4(4) (2014) 1139-1143.
9) R. Watanabe, N. Hirata, K. Miura, Y. Yoda, Y. Fushimi, C. Fukuhara, *Applied Catalysis A: General*, 587 (2019) 117238.
10) R. Watanabe, N. Hirata, Y. Yoda, Y. Fushimi, C. Fukuhara, *Journal of the Japan Petroleum Institute*, 63(4) (2020) 228-237.

An engineered Bimetallic Fe-Cu-MOR Zeolite Catalyst for Direct Oxidation of Methane to Methanol

Ijaz Hussain [a], Khalid Alhooshani [a, b *], Saheed A Ganiyu [a, b], Hassan Alasiri [a, c].

[a] Center for Refining & Advanced Chemicals, King Fahd University of Petroleum & Minerals, Dhahran, 31261, Saudi Arabia

[b] Department of Chemical Engineering, King Fahd University of Petroleum & Minerals, Dhahran, 31261, Saudi Arabia

[c] Department of Chemistry, King Fahd University of Petroleum and Minerals, Dhahran 31261, Saudi Arabia

Over the past few decades, the overuse of fossil fuels has contributed to disastrous climate change. Methane is one of the most potent greenhouse gases and must be controlled immediately to combat climate change. According to the waste-to-wealth strategy, oxidizing methane to produce methanol is a promising approach to reducing greenhouse gas emissions and limiting climate change. In the present study, a bimetallic Fe-Cu-MOR zeolite catalyst was fabricated to directly oxidize methane to methanol using an incipient wet impregnation method. Different techniques were used to characterize the synthesized catalyst, such as XRD, N_2 adsorption analysis, and FTIR spectroscopy. The direct oxidation of methane to methanol shows a higher activity over Cu-Fe-MOR with 5% Fe loading. The synthesized Cu-Fe-MOR catalyst produced a high yield of methanol (60 mol/$g^{-1}{}_{cat}$) compared to other catalysts of 1 and 3% loading of Fe. Based on the observations, it is possible to conclude that Cu-Fe-MOR has significant potential to produce methanol, both industrially and academically.

CO_2-assisted Dehydrogenation of Propane to Produce Propylene and Ethylene using Bimetallic Catalysts

Emad N. Al-Shafei [*]
[*] Research and Development Center, Saudi Aramco, Dhahran, Saudi Arabia

Abstract

Given major increases in the demand for light olefins, especially in favor of propylene, the growth of on-purpose propylene production technologies is significantly accelerating with increase in global production. Existing on-purpose propane dehydrogenation (PDH) plants are mainly equipped with chromium or platinum-based catalysts and operate at low hourly space velocity. As such, they limited to propylene yields and suffer from low time-on-stream where catalyst deactivation could be observed within hours. In this study, non-oxidative catalysts developed for the utilization of CO_2 with the aim of producing greater yields of both propylene and ethylene, to improve the techno-economic competitiveness of the PDH process. The study revealed that bimetallic catalyst, supported by titanium oxide, show promise in increasing olefins yields, while operating under conventional PDH operating conditions. In the presence of CO_2, PDH developed catalysts resulted in an increase of the catalytic stability and conversion for longer time-on-stream (30-250 hours) with high olefin yield of over 40%. The study revealed the important of titania-based catalyst in promoting the C-H and C-C bond activation of propane and demonstrated an alternative process for the production of stable yields of propylene and ethylene.

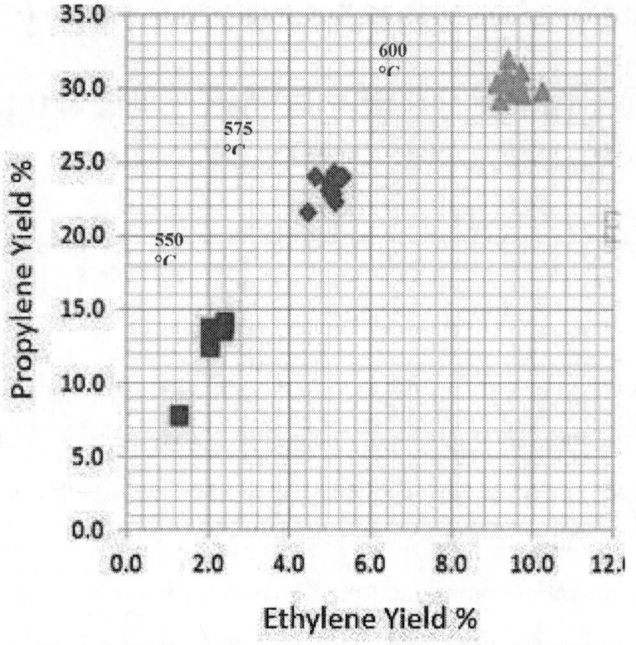

Total yield of propene and ethene of 550, 575 and 600 °C

CO_2 capture and conversion analysis technology development: TRL 1 -to- TRL 9

Ali S. Al-Hunaidy[1], Hasan Imran[1] , Wonsuk Chung[2], Jay H. Lee[2]

1 Carbon Management Division, Research & Development Center, Saudi Aramco, Dhahran 31311, Kingdom of Saudi Arabia

2 Department of Chemical and Biomolecular Engineering, Korea Advanced Institute of Science and Technology (KAIST), 291 Daehak-ro, Yuseong-gu, Daejeon 34141, Republic of Korea.

Climate change concern keeps arising with initiatives kicked off to achieve net-zero emissions through mandating the implementation of different CO_2 reduction strategies including CO_2 utilization and storage. However, the challenge is always anticipated at the capture cost and conversion of CO_2. In addition, nowadays, large number of CO_2 conversion technologies are being considered for potential commercial use. Therefore, it is essential to have a reliable tool that can assess and select the best conversion technology based on technical and economic viability.

A computer-aided tool, ArkaTAC[3] (Aramco-KAIST tool for analyzing CO_2 capture and conversion technologies), for sustainability identification and optimization of CCU systems has been developed. The tool can analyze a CCU superstructure with multiple carbon sources, capture processes, conversion technologies, and potential utilization products by mathematical programming. The life cycle inventory (LCI) database in the tool allows precise evaluation of the CCU system for a specific scenario. A graphic user interface of the tool visualizes the sustainability results and helps the users understand the CCU system.

The industry including oil refineries and gas plants, academia and policy makers can use the software tool to screen relevant CO_2 capture, storage and conversion options and identify the most appropriate reduction solutions and their economic impacts in a reliable and efficient manner.

Software development steps from TRL 1 started with knowledge development and passing through the technology development while ending up with business development at TRL 9 will be presented. Software live demonstration will be conducted as well.

Ceria-catalyzed synthesis of aliphatic polycarbonate diols from atmospheric flow carbon dioxide and diols without using dehydrating agents

Keiichi Tomishige

Department of Applied Chemistry, School of Engineering, Tohoku University, Aoba 6-6-07, Aramaki, Aoba-ku, Sendai 980-8579, Japan

E mail: tomishige@tohoku.ac.jp

Abstract

Much more attention has been recently paid to the conversion of CO_2 to useful chemicals from the viewpoint of carbon recycling, carbon neutrality and so on. The direct transformation of CO_2 and diols into aliphatic polycarbonate diols is promising as an alternative method for the phosgene process, which can decrease the utilization of phosgene. However, the direct transformation of CO_2 and diols into aliphatic polycarbonate diols is so challenging due to the inert characteristic of CO_2 and the severe limitation from the thermodynamics of the reactions. The presentation includes the direct synthesis of polycarbonate diols from atmospheric pressure CO_2 and α,ω-diols using CeO_2 catalyst and CO_2 flow semi-batch reactor. The target alternating polycarbonate diol from CO_2 and 1,6-hexanediol was obtained with high yield and selectivity without using any dehydrating agents. Activation of atmospheric pressure CO_2 by a CeO_2 catalyst and the shift of equilibrium towards the product by removing the coproduced water (gas stripping) are responsible for the high yield.

1. Introduction

Conversion of CO_2 into useful compounds can contribute to the construction of future sustainable society, carbon neutrality, carbon recycling, and so on. A variety of CO_2 conversion technologies can be divided into two categories: reductive conversion and non-reductive conversion. The reductive conversion includes the production of hydrocarbons and oxygenates from CO_2 by the reduction with H_2 and artificial photosynthesis. On the other hand, compounds such as ureas, carbamates, and carbonates are also useful, and the oxidation state of the carbon atom in these compounds is the same as that in CO_2 molecule. Therefore, the synthesis of ureas, carbamates and carbonates from CO_2 and corresponding alcohols and/or amines is regarded as non-reductive CO_2 conversion, where the impact of the non-reductive conversion is to substitute phosgene with CO_2 [1-10]. Phosgene has very high reactivity, and the reaction of phosgene with alcohols and amines can proceed without using catalysts. In contrast, the reactivity of CO_2 is low and a suitable catalyst is needed in the reaction of CO_2 with alcohols and amines. Our group has reported that CeO_2 can catalyze the reaction of CO_2 with alcohols giving corresponding carbonates selectively [11, 12]. It has been known that the yield of the carbonates is limited by the reaction equilibrium [12-14]. It has also been reported that the combination of the reaction of CO_2 and alcohols with the hydration of nitriles enabled the higher yield of the corresponding carbonates [15-22].

The target products in my presentation are aliphatic polycarbonate diols directly synthesized from CO_2 and α,ω-diols. The conventional phosgene method for the synthesis of the aliphatic polycarbonates is shown below. In the reaction equation, 1,4-butanediol, 1,5-pentanediol, and 1,6-hexanediol are applicable, representing $m = 1$ to 3.

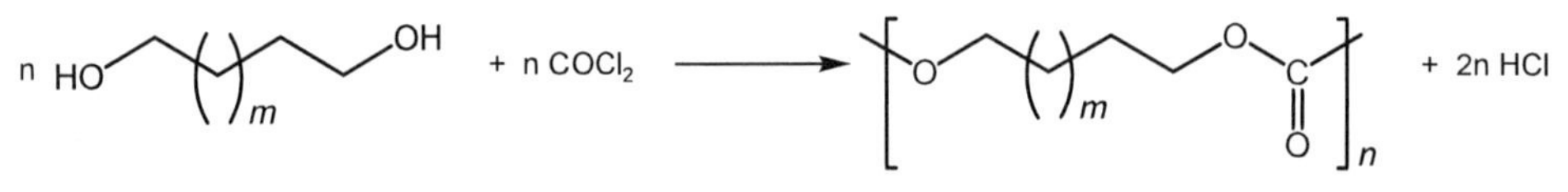

A variety of the methods for the synthesis of aliphatic polycarbonates from CO_2 and α,ω-diols have been proposed. The synthesis of polycarbonates from α,ω-diols, dihalides, CO_2 and Cs_2CO_3 has been reported as shown below [23, 24].

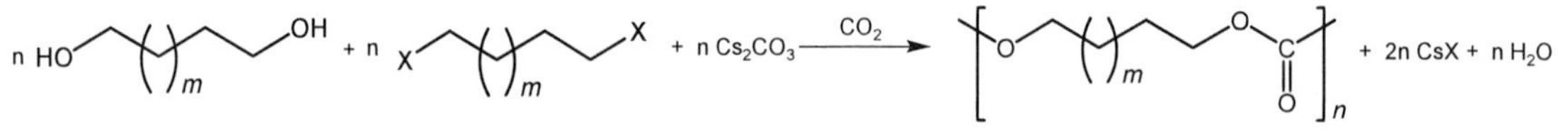

Another method is the synthesis of polycarbonates from α,ω-diols, CH_2X_2, CO_2 and Cs_2CO_3 as shown below [25, 26]. The reaction can procced even under rather low CO_2 pressure (~ 0.1 MPa). Another important point is that the ether bond derived from CH_2X_2 was formed together with the carbonate bond.

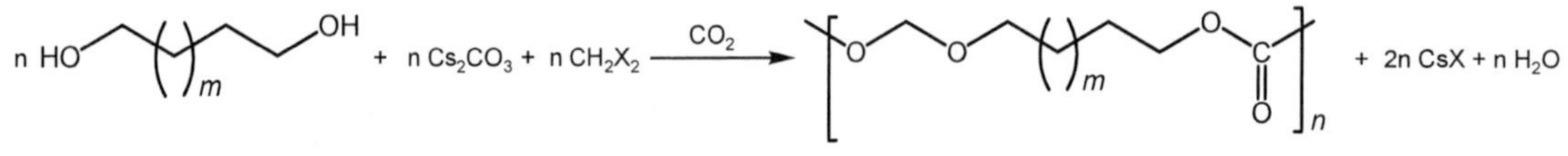

Aliphatic polycarbonates can be also synthesized from α,ω-diols, CX_4, CO_2, trialkylphosphine and a base such as *N*-cyclohexyl-*N',N',N'',N''*-tetramethylguanidine (CyTMG) as shown below [27, 28]. In this method, it is possible to synthesize polycarbonate from ethylene glycol.

It is common that the synthesis methods listed above utilize the non-catalytic and stoichiometric reactions and use toxic organic halides. The catalytic synthesis methods have been recently reported. The direct reaction of α,ω-diols with CO_2 by the combination of the hydration of nitriles (2-cyanopyridine or 2-furoniritle), where both reactions can be catalyzed by CeO_2 as a heterogeneous catalyst [29-32].

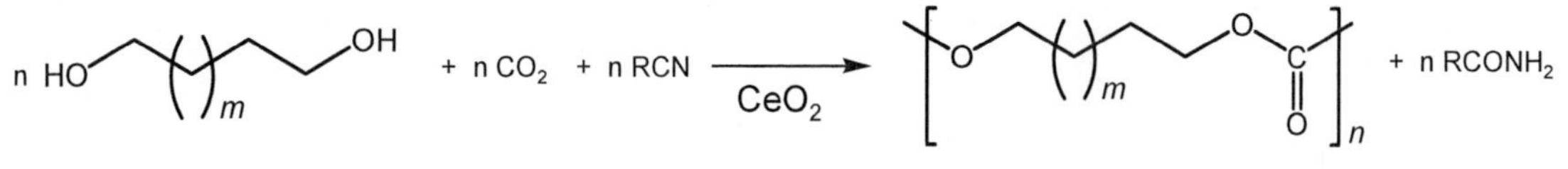

The boiling points of diol substrates and the product in the synthesis of aliphatic polycarbonates are clearly higher than that of H_2O. This situation is different from the case of the synthesis of simple carbonates such as dimethyl carbonate from volatile alcohols. Therefore, it is expected that a high yield of target polycarbonate can be obtained by selectively removing the coproduced water from the reaction media by CO_2 flow and shifting the equilibrium to the product side if diols can react with atmospheric pressure CO_2. Here, I introduce that the combination of CeO_2 catalyst and CO_2 flow reaction setup was an effective reaction system for the direct synthesis of polycarbonate diols from atmospheric CO_2 and diols, and the reaction equation is shown below [33].

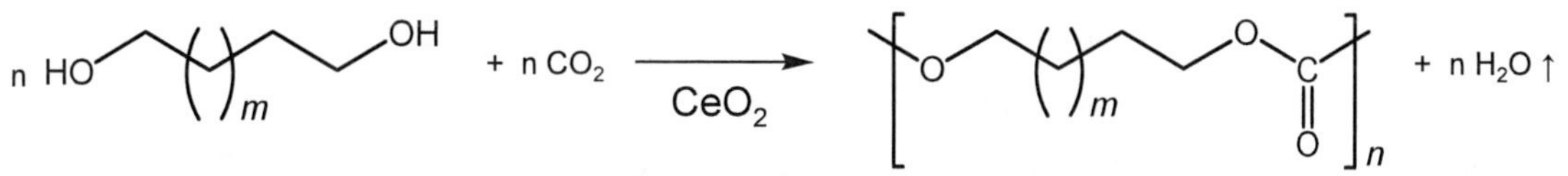

2. Experimental [33]

General procedure for the reaction of 1,6-hexanediol and CO_2 in the CO_2 flow semi-batch reactor: The reaction was carried out in a set of glass apparatus including a 50 mL 3-neck pear-shaped flask at bottom, a Grahams condenser (under ambient temperature) at the middle, and a glass trap filled with ethanol on the top (Figure 1). A catalyst and reagents were put into the flask, and then a PTFE tube for CO_2 introduction was inserted to the bottom of the flask. The glass apparatus was finally set up, and the flask part was put into an oil bath for heat-up. After an appropriate reaction time, the flask was transferred to an ice-water bath for cooling down. The ethanol in the glass trap was firstly collected with 1 g of diglyme (internal standard) for GC analysis. After the flask was cooled down to room temperature, 40 g of THF and 1 g of diglyme as an internal standard were injected, and the mixture was collected for further analysis.

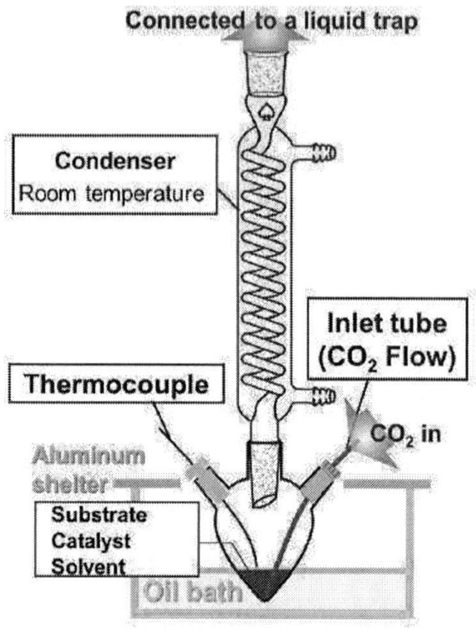

Figure 1. Semi-batch reactor setup for the synthesis of aliphatic polycarbonate diols from α,ω-diols and atmospheric flow CO_2 using CeO_2 catalyst [33].

Preparation of CeO_2 catalyst was carried out by calcining CeO_2-HS (Daiichi Kigenso, Japan. The purity of CeO_2: 99.97%) for 3 h in air at 873 K. The specific surface area (BET method) of the CeO_2 was 84 m^2/g.

3. Results and discussion [33]

Figure 2 shows the result of the reaction of 1,6-hexanediol with CO_2 without using a catalyst and with various metal oxides. No products were detected without using a catalyst. CeO_2 exhibited excellent catalytic activity towards the reaction of flow CO_2 and 1,6-hexanediol under atmospheric pressure, giving 17% conversion and high selectivity to the carbonates (dimer 89%, polymer 11%) of 1,6-hexanediol. In the cases of γ-Al_2O_3 and SiO_2-Al_2O_3, there were no carbonate products, and linear and cyclic ethers of 1,6-hexanediol were produced as main products with minor products of 5-hexen-1-ol and its isomers, which can be formed by the acidity of these catalysts. Other metal oxides used in the experiments were inactive, and no products were detected at all.

Figure 3 shows the effect of CeO_2 amount in the reaction of 1,6-hexanediol with CO_2. The conversion at 1 h was plotted as a function of CeO_2 amount, showing that the conversion increased with the increase of CeO_2 amount in the range of 0-0.08 g and reached 3.8% in this range, where the conversion was controlled by the amount of CeO_2, indicating that the rate-determining step can be the catalytic reaction on CeO_2. With a larger amount of CeO_2 from 0.08 to 0.20 g, the conversion was almost constant at 3.8%, indicating that the rate-determining step is the removal of water from the reactor.

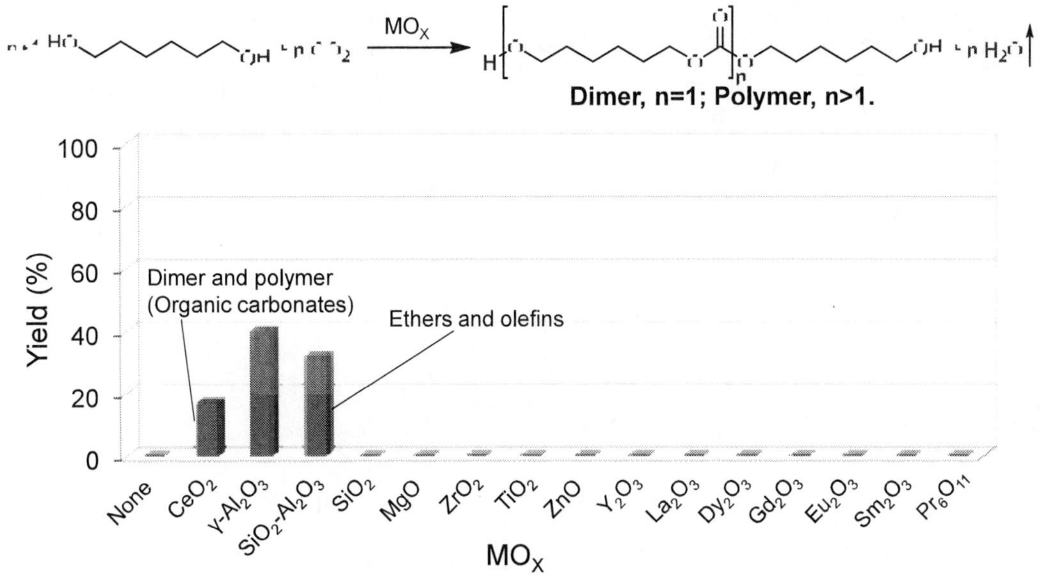

Figure 2. Result of the reaction of 1,6-hexanediol with CO_2 without using catalyst and with various metal oxides.

Reaction conditions: MO_X 0.10 g, 1,6-hexanediol 2.0 g, triglyme 2 g, diphenyl ether 3 g, 483 K, CO_2 flow 200 mL min^{-1}, 4 h.

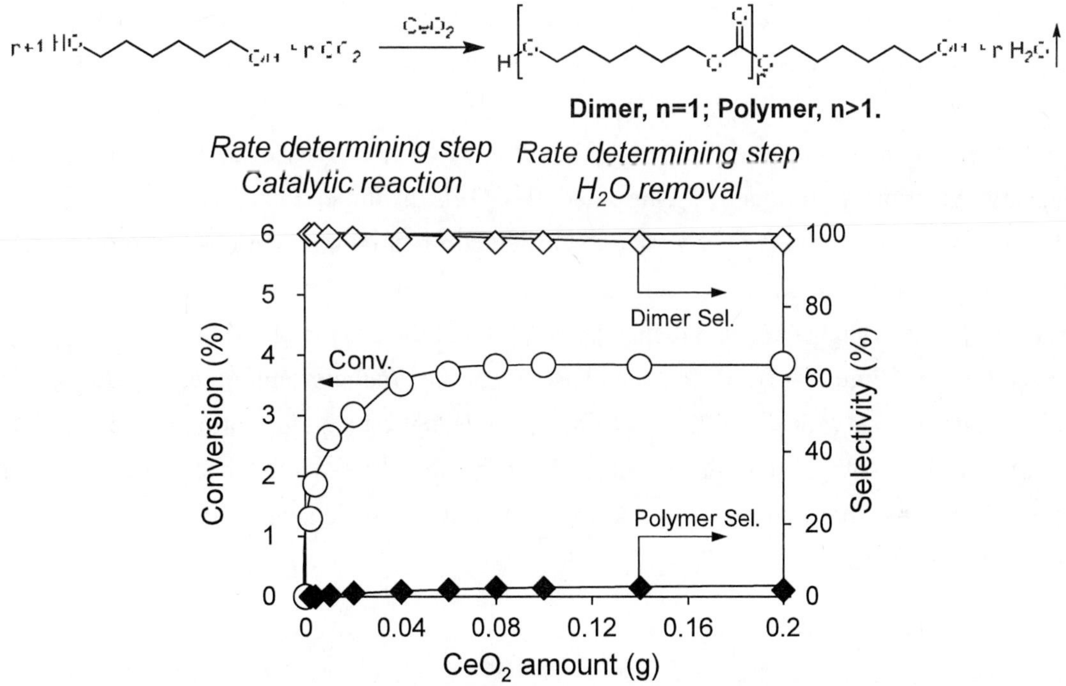

Figure 3. Effect of the amount of CeO_2 in the reaction of 1,6-hexanediol with CO_2

Reaction conditions: CeO_2 0-0.20 g, 1,6-hexanediol 2.0 g (17 mmol), triglyme 3 g, CO_2 flow rate 200 mL min^{-1}, 473 K, 1 h.

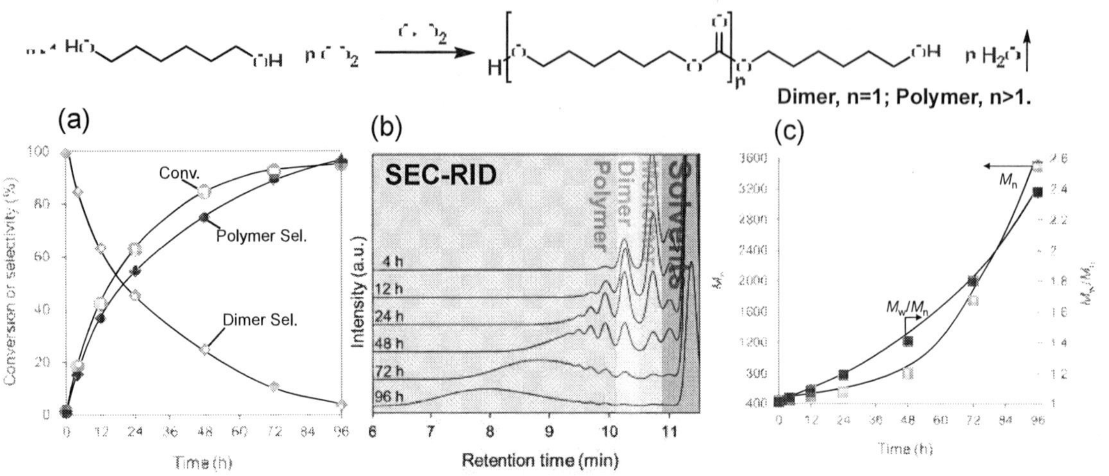

Figure 4. Results of the reaction of 1,6-hexanediol with CO_2 on CeO_2.
(a) Conversion and selectivity as a function of the reaction time
(b) SEC-RID of the samples
(c) Molecular weight (M_n) and dispersity (M_w/M_n) profile
Reaction conditions: CeO_2 0.10 g, 1,6-hexanediol 2.0 g (17 mmol), triglyme 2 g, diphenyl ether 3 g, CO_2 flow rate 200 mL min^{-1}, 483 K, 0-96 h.

Figure 4(a) shows the conversion and selectivity as a function of the reaction time in the reaction of 1,6-hexanediol with CO_2 on CeO_2. The reaction of 1,6-hexanediol proceeded smoothly to reach 95% conversion at 96 h. The selectivity to the dimer was high at a low conversion level and decreased with longer reaction time, and conversely, the selectivity to the polymer kept increasing and reached 97% at 96 h. High selectivity of the dimer product means that the primary product of 1,2-hexanediol+CO_2 is the dimeric carbonate (bis(6-hydroxyhexyl) carbonate). The size-exclusion chromatography (SEC)-RID profiles are shown in Figure 4(b), which clearly demonstrated the formation of polycarbonate. The average molecular weight (M_n) and dispersity (M_w/M_n) of the produced products were analyzed by SEC (Figure 4(c)). Both M_n and M_w/M_n of the products kept increasing during the reaction, and polymer with M_n of 3500 obtained at 96 h suggests the formation of the polycarbonate diol with 24 repeating units. In addition, the increase of M_n became faster when higher conversions were achieved with longer reaction times, which can be due to step-growth polymerization.

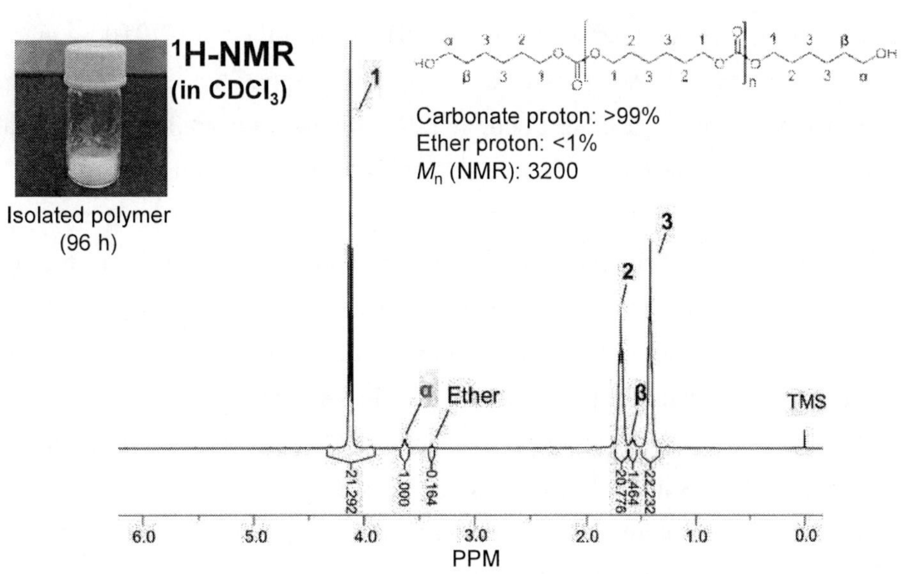

Figure 5. ^{1}H-NMR of the isolated polymer formed at 96 h in Figure 4.

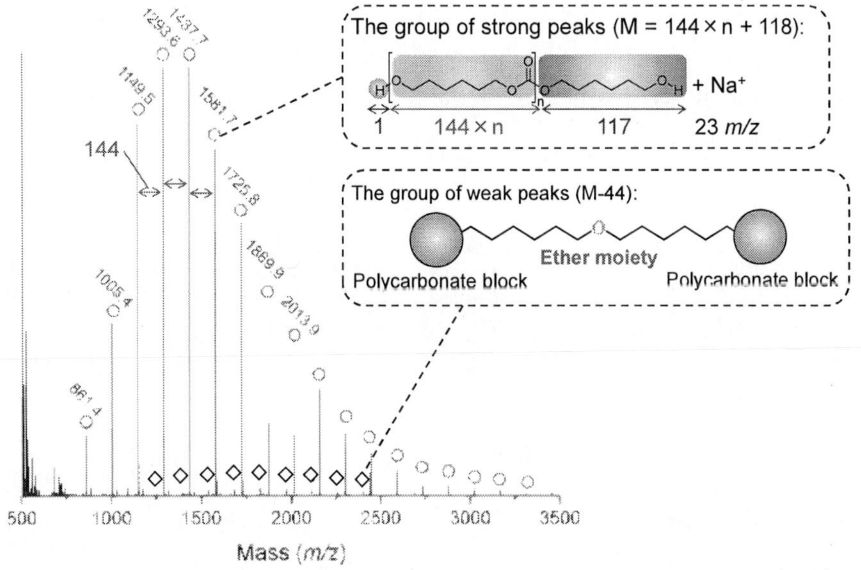

Figure 6. MALDI-TOF-MS data of the isolated polymer formed at 96 h in Figure 4.

The polymer produced at 96 h was isolated by precipitation, and a pure white wax-like solid (photo in Figure 5) was obtained. The isolated polymer was analyzed by ^{1}H-NMR (Figure 5) and MALDI-TOF-MS (Figure 6). The signals at 3.6 ppm and 4.0-4.2 ppm in Figure 5 can be assigned to the protons of the methylene group adjacent to the OH group and carbonate group, respectively. The average molecular weight of the isolated polycarbonate diol was calculated to be 3200 g mol^{-1} on the basis of the ratio of the proton numbers on the two positions ("α" and "1" in Figure 5). The obtained value agreed with the result of SEC (3500 g mol^{-1} based on polystyrene standards). A triplet signal at 3.4 ppm is assigned to the protons of the methylene group adjacent to the ether bond, and the ratio of ether linkage to carbonate linkage is estimated to be as small as 0.8%. This value also agrees well with the result of MALDI-TOF-MS (Figure 6), where a series of strong signals with a mass increment of 144 can be assigned to poly(hexamethylene carbonate) diol and the other series of minor signals with 44 lower mass number than the strong signals can be assigned to poly(hexamethylene carbonate) diol with one ether linkage replacing one carbonate linkage. In the previous reports, very high selectivity to organic carbonates and very low selectivity to ethers in the reaction of CO_2 with alcohols have been observed on CeO_2 catalyst [12]; however, an important point is that the reaction conditions in the present results are clearly different from the previous works from the aspect of CO_2 pressure and reaction temperature. High selectivity to the carbonate linkage in the present result indicates that CeO_2 exhibited high selectivity to carbonate formation even at low CO_2 pressure and high reaction temperature.

Figure 7 shows the reaction scheme of the formation of the carbonate dimer and polycarbonate diol from 1,6-hexanediol and CO_2 over CeO_2. At the first step, the formation of the carbonate dimer is selectively at the initial stage, and this suggests that the reactivity of 1,6-hexanediol is much higher than that of the produced carbonate dimer. Consecutive carbonate formation from the dimer, 1,6-hexanediol and CO_2 can be connected to the formation of the polycarbonate diol.

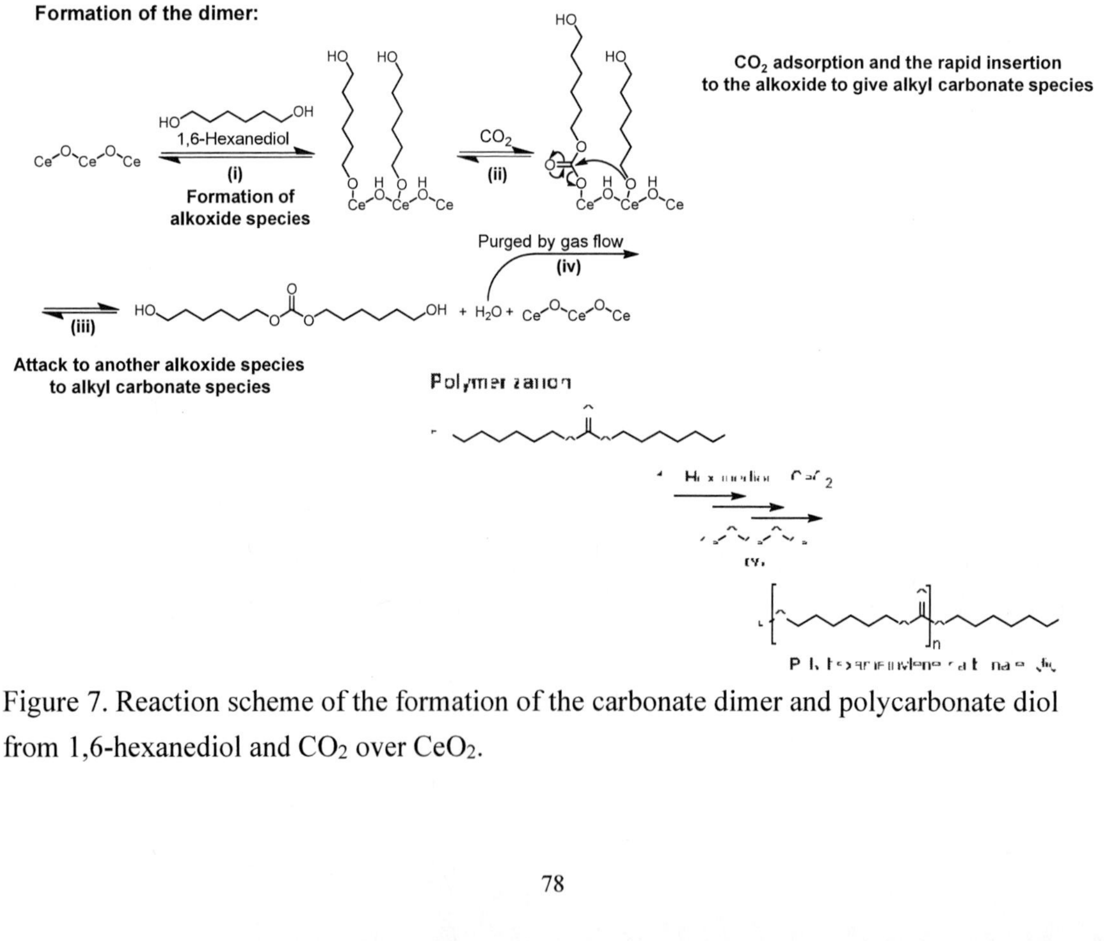

Figure 7. Reaction scheme of the formation of the carbonate dimer and polycarbonate diol from 1,6-hexanediol and CO_2 over CeO_2.

4. Conclusion

Direct conversion of atmospheric pressure flow CO_2 with 1,6-hexanediol to the polycarbonate diol without using any dehydrating agents was demonstrated by using a heterogeneous CeO_2 catalyst and the semi-batch reactor. The alternating polycarbonate diol without almost no ether linkage is selectively formed, and the yield is also high.

5. References

[1] K. Tomishige, M. Tamura, Y. Nakagawa, Chem. Rec. 19, 1354 (2019).

[2] M. Honda, S. Sonehara, H. Yasuda, Y. Nakagawa, K. Tomishige, Green Chem., 13, 3406 (2011).

[3] M. Tamura, K. Noro, M. Honda, Y. Nakagawa, K. Tomishige, Green Chem., 15, 1567 (2013).

[4] M. Tamura, M. Honda, Y. Nakagawa, K. Tomishige J. Chem. Technol. Biotechnol., 89, 19 (2014).

[5] M. Tamura, K. Ito, Y. Nakagawa, K. Tomishige, J. Catal., 343, 75 (2016).

[6] M. Tamura, M. Honda, K. Noro, Y. Nakagawa, K. Tomishige, J. Catal., 305, 191 (2013).

[7] M. Tamura, A. Miura, M. Honda, Y. Gu, Y. Nakagawa, K. Tomishige, ChemCatChem, 10, 4821 (2018).

[8] Y. Gu, A. Miura, M. Tamura, Y. Nakagawa, K. Tomishige, ACS Sustainable Chem. Eng., 7, 16795 (2019).

[9] J. Peng, M. Tamura, M. Yabushita, R. Fujii, Y. Nakagawa, K. Tomishige, ACS Omega, 6, 27527 (2021) .

[10] J. Peng, M. Yabushita, Y. Li, R. Fujii, M. Tamura, Y. Nakagawa, K. Tomishige, Appl. Catal. A, 643, 118747 (2022).

[11] K. Tomishige, Y. Gu, M. Tamura, Y. Nakagawa, Materials Today Sustainability, 9, 100035 (2020).

[12] Y. Yoshida, Y. Arai, S. Kado, K. Kunimori, K. Tomishige, Catal. Today, 115, 95 (2006).

[13] T. Chang, M. Tamura, Y. Nakagawa, N. Fukaya, J. Choi, T. Mishima, S. Matsumoto, S. Hamura, K. Tomishige, Green Chem., 22, 7321 (2020).

[14] K. Tomishige, Y. Gu, Y. Nakagawa, M. Tamura, Frontiers in Energy Research, 8, 117 (2020).

[15] M. Honda, M. Tamura, Y. Nakagawa, K. Tomishige, Catal. Sci. Technol., 4, 2830 (2014).

[16] M. Honda, A. Suzuki, N. Begum, K. Fujimoto, K. Suzuki, K. Tomishige, Chem. Commun., 4596 (2009).

[17] M. Honda, S. Kuno, N. Begum, K. Fujimoto, K. Suzuki, Y. Nakagawa, K. Tomishige, Appl. Catal. A, 384, 165 (2010).

[18] M. Honda, S. Kuno, S. Sonehara, K. Fujimoto, K. Suzuki, Y. Nakagawa, K. Tomishige,

ChemCatChem, 3, 365 (2011).
[19] M. Honda, M. Tamura, Y. Nakagawa, S. Sonehara, K. Suzuki, K. Fujimoto, K. Tomishige, ChemSusChem, 6, 1341 (2013).
[20] M. Honda, M. Tamura, Y. Nakagawa, K. Nakao, K. Suzuki, K. Tomishige, J. Catal., 318, 95 (2014).
[21] M. Honda, M. Tamura, K. Nakao, K. Suzuki, Y. Nakagawa, K. Tomishige, ACS Catal., 4, 1893 (2014).
[22] H. Ohno, M. Ikhlayel, M. Tamura, K. Nakao, K. Suzuki, K. Morita, Y. Kato, K. Tomishige, Y. Fukushima, Green Chem., 23, 457 (2021).
[23] S. Oi, K. Nemoto, S. Matsuno, Y. Inoue, Macromol. Rapid Commun., 15, 133 (1994).
[24] Z. Chen, N. Hadjichristidis, X. Feng, Y. Gnanou, Polym. Chem., 7, 4944 (2016).
[25] S. Bian, C. Pagan, A. A. A. "Artemyeva", G. Du, ACS Omega, 1, 1049 (2016).
[26] S. Bian, A. A. Andrianova, A. Kubatova, G. Du, Mater. Today Commun., 18, 100 (2019).
[27] J. Kadokawa, H. Habu, S. Fukamachi, M. Karasu, H. Tagaya, K. Chiba, Macromol. Rapid Commun., 19, 657 (1998).
[28] J. Kadokawa, S. Fukamachi, H. Tagaya, K. Chiba, Polym. J., 32, 703 (2000).
[29] M. Tamura, K. Ito, M. Honda, Y. Nakagawa, H. Sugimoto, K. Tomishige, Sci. Rep., 6, 24038 (2016).
[30] Z. Gong, Y. Li, H. Wu, S. D. Lin, W. Yu, Appl. Catal. B, 265, 1185 (2020).
[31] Y. Yu, T. Wang, L. Chang, P. Wu, B. Yu, W. Yu, J. Taiwan Inst. Chem. Eng., 116, 36 (2020).
[32] Y. Gu, K. Matsuda, A. Nakayama, M. Tamura, Y. Nakagawa, K. Tomishige, ACS Sustain. Chem. Eng., 7, 6304 (2019).
[33] Y. Gu, M. Tamura, Y. Nakagawa, K. Nakao, K. Suzuki, K. Tomishige, Green Chem., 23, 5786 (2021).

Contaminants in plastic derived oil: A thorn in the flesh of plastic circularity?

Aaron Akah, Hendrik Muller, Enas Altarooti, Rashed Aleisa and Rasha Daadoush

R&D Center, Saudi Aramco

This presentation covers the impacts of contaminants that are prevalent in plastic waste derived oil (PDO) and looks at strategies to mitigate the harmful effects of these contaminants. Feedstock recycling of plastic waste via pyrolysis provides a potential route for the conversion of plastic waste to useful petrochemicals (circular chemicals), via integration with steam crackers or refineries. However, the huge variation in the composition of plastic waste leads to PDO with different qualities, especially with regards to the contaminants. This makes it difficult to predict the quality of PDO. Analysis of plastic derived waste oils shows that they contain a higher concentration, and very different types of contaminants compared to traditional hydrocarbon-based feedstocks. As most refineries and petrochemical facilities are not used to the levels of contaminants found in PDO, they are uncertain about integrating PDO into their installations. Therefore, comprehensive analysis is required to understand the levels and types of contaminants, that each PDO contains. Analyses show that the main contaminants in PDO are molecules containing heteroatoms (chlorine, nitrogen, oxygen, sulfur), unsaturated molecules, and metals (Ca, Fe, Zn, Na, Si, etc). The results have led to the evaluation of techniques for decontamination and subsequent conversion of PDO oil to circular chemicals.

Conversion of rubber waste to liquid hydrocarbon fuel via thermal liquefaction process using solvent

Nabeel Ahmad
Center for Refining and Advanced Chemicals, King Fahd University of Petroleum and Minerals, Dhahran 31261, Saudi Arabia

(E-mail: nabeel.ahmad@kfupm.edu.sa)

ABSTRACT

Natural rubber (NR) was converted to liquid hydrocarbon fuel in the presence of methyl alcohol (CH_3OH) using autoclave reactor. Different process parameters were assessed for their effect on product quantity and quality. At optimum condition, the maximum amount of oil obtained was ~80wt%. GCMS analysis indicated the presence of aromatics, alkyls, aldehydes, and ketones in oil.

Keywords: Rubber waste, liquid-oil, thermal liquefaction, methyl alcohol.

INTRODUCTION

Among different alternative energy sources, NR waste is also one of the potential sources that can contribute to upcoming energy demand. NR waste can be transformed into liquid fuels and high value-added chemical by thermal conversion methods. Chemically, NR is the polymer of isoprene. NR is an important feedstock used in different industries in huge quantities. It is tropical planation crops available in south Asian countries and African countries. According to statistical data, the average global production of NR was estimated to be 17,338 thousand tonnes from the years 2016-2020. Moreover, during COVID-19 pandemic, the utilization of NR was increased drastically above 65 billion single-use in the form of rubber gloves globally. This extensive utilization of NR has led to an intensive upsurge in rubber-based wastes, whereas the recycling rate is only ~1.5%. In order to attain the carbon neutrality, the existing rubber recycling methodologies like burning and landfilling are becoming increasingly unfeasible. Therefore, chemical upcycling of R1 to synthesize fuels would be progressive approach toward sustainable rubber and energy industry.

NR was characterized using TGA and elemental analysis to understand its thermal and chemical properties, then it was liquefied at low temperature using thermal liquefaction technique to produce liquid fuels using CH_3OH. The effect of different parameters such as temperature, material to solvent ratio and time were studied.

MATERIALS

NR of grade Ribbed Smoked Sheets (RSS) was used in this study was sliced into particle size of about 0.1-0.4 cm before its usage in the experimentation. Methyl alcohol used was obtained from Sigma Aldrich. Inert N_2 gas was used to purge an autoclave reactor.

METHODOLOGY

The conversion of NR to liquid hydrocarbon fuel was performed in autoclave reactor. All the process parameters, settings, and separation techniques are shown in Figure 1.

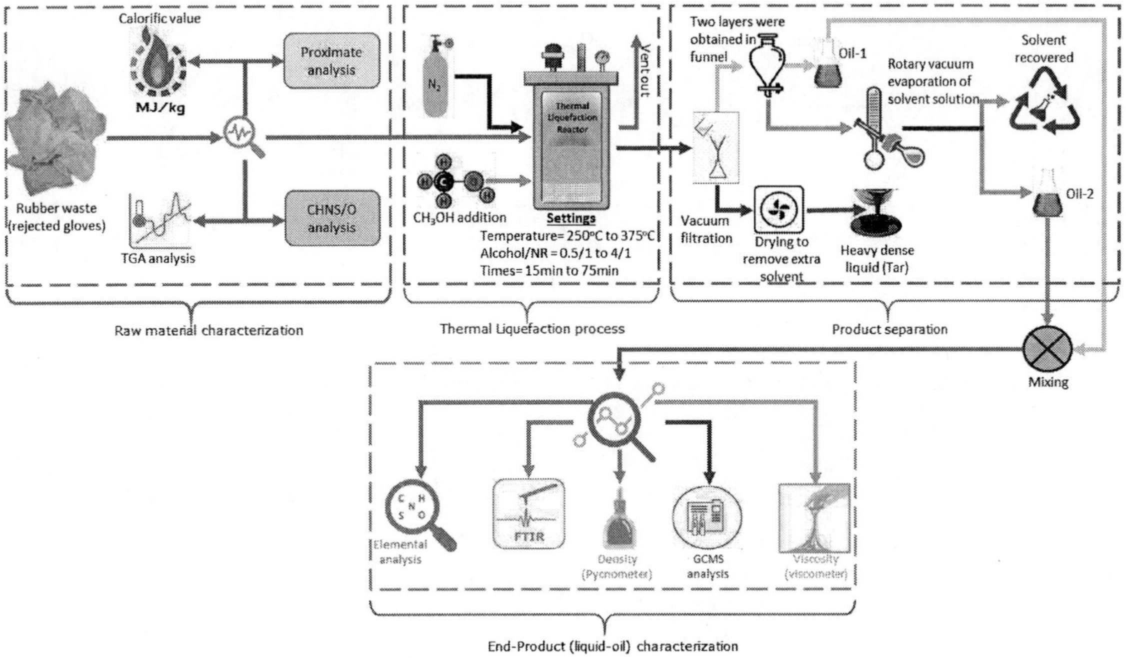

Figure 1. Process flow and schematic diagram thermal liquefaction process

The yield of oil, tar and gases were projected using eq (1) & (2). While, the yield of gases and light end losses were calculated using equation (2).

$$\text{Yield}(\%) = \frac{\text{weight. of product}}{\text{weight. of rubber+weight. of methanol consumed}} \times 100 \qquad (1)$$

$$\text{Gas yield } (\%) = 100 - (\text{oil}\% + \text{Tar}\%) \qquad (2)$$

After the separation of final product, it was sent for characterization using different analytical techniques such as FTIR, elemental analysis, GCMS etc.

RESULTS

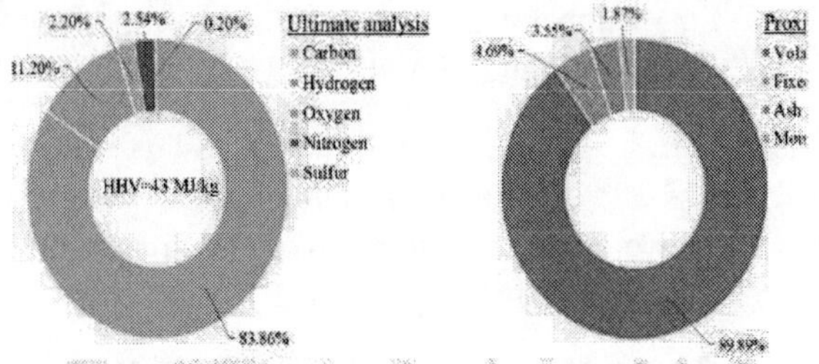

Figure 2. Ultimate and proximate analysis of feedstock

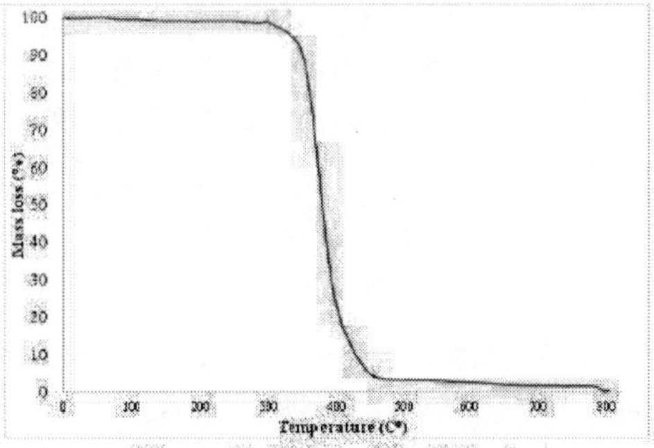

Figure 3. TGA of feedstock

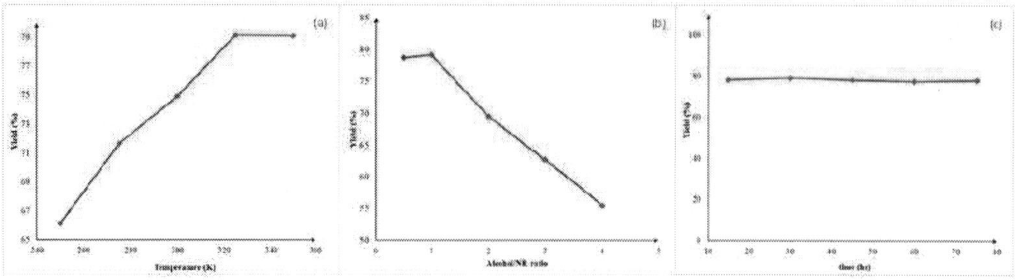

Figure 5. Effect of parameters on oil yield; (a) temperature, (b) alcohol/NR ratio; & (c) time

Table 1. Properties of oil at optimum conditions

Properties of oil	
Carbon	86.11%
Hydrogen	12.5%
Nitrogen	0.41%
Oxygen	0.77%
Sulfur	0.21%
HHV(MJ/kg)	46.36%
Density (g/ml)	88.00%
Viscosity (cP)	11.70%

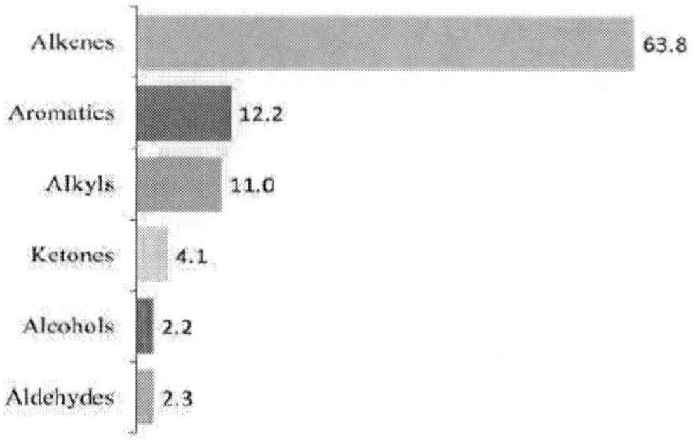

Figure 4. Families of compounds present in oil (via GCMS)

CONCLUSION: At optimum conditions, approximately ~80wt % of liquid hydrocarbon fuel was obtained from the liquefaction of waste rubber in the presence of methyl alcohol solvent. The obtained oil exhibits good fuel properties in terms of heating value, low sulfur.

REFERENCES:

1. N. Ahmad, F. Abnisa and W.M.A.W. Daud, Potential use of natural rubber to produce liquid fuels using hydrous pyrolysis - A review, RSC Advances 6 (2016) 68906. http://dx.doi.org/10.1039/C6RA09085K

2. H. Chittella, L.W. Yoon, S. Ramarad and Z.-W. Lai, Rubber waste management: A review on methods, mechanism, and prospects, Polymer Degradation and Stability 194 (2021) 109761. https://doi.org/10.1016/j.polymdegradstab.2021.109761

3. J. Wang, J. Jiang, X. Meng, S.S. Lam, A.J. Ragauskas and Y. Wang, Catalytic cascade upcycling single-use natural rubber glove wastes into fuels via a two-stage pressurized fixed-bed reactor, Fuel Processing Technology 238 (2022) 107490. https://doi.org/10.1016/j.fuproc.2022.107490

4. N. Ahmad, F. Abnisa and W.M.A. Wan Daud, Liquefaction of natural rubber to liquid fuels via hydrous pyrolysis, Fuel 218 (2018) 227. https://doi.org/10.1016/j.fuel.2017.12.117

Chemometrics Study of Sulfur Reactivity in Hydrodesulfurization Process

Hutoon Alhargan and Hendrik Muller
Saudi Aramco, Research & Development Center, 31311, Dhahran, Saudi Arabia
hutoon.alhargan@aramco.com

A demetallized, i.e., de-asphalted vacuum residue (DMO) of an Arabian crude oil was subjected to progressive hydrotreating conditions to reduce its total sulfur content. The DMO composition was monitored using Fourier Transform – Ion Cyclotron Resonance Mass Spectrometry (FT-ICR MS) in conjunction with Atmospheric Pressure Photo Ionization (APPI) operating in positive polarity. The spectra generated revealed a vast number of data points for each sample, related to hydrocarbon, sulfur, and other heteroatom containing compounds and their distribution across carbon number and DBE. The entire data set from high to low sulfur content was processed using a multivariate method; Principal Component Analysis (PCA). PCA captures the covariation and correlation of a wide range of variables between the different data sets, thereby ascertaining the most significant, underlying variables to trace the trajectory of systematic changes in the sulfur aromatic species with progressive reduction of the total sulfur content. The score and loading of the principle components were linked to the underlying chemical features, which revealed the reactivity of the DMO components. While the HDS reactivity of lighter fractions, such as gas oils, is well established, the presented results confirm that the HDS reactivity are similar even in a very heavy DMO fraction. In addition, the analysis reveals that heavy polysulfur species, which are found only in the DMO, are swiftly removed, while alkyl-substituted dibenzothiophenes remain as the most recalcitrant sulfur compounds. The presented work shows that the developed methodology can be applied to ascertain novel reactivities of hitherto inaccessible, non-boiling petroleum fractions and tailor future catalysts to target the recalcitrant sulfur species in the DMO.

Development of Hierarchical CoMo/Beta Zeolite for BTX Production

Eman Z. Albaher

Saudi Aramco R&DC

Hierarchical Beta zeolites were synthesized with different Si/Al ratios and with different NaOH concentrations which were synthesized through surfactant templated approach. Cobalt with molybdenum on H-Beta zeolite catalysts were prepared through co-impregnation method with the same metallic loadings. The in-house synthesized materials were compared with commercial zeolites. The activity of the catalysts was determined for selective hydrocracking of a model compound which is 1-methylnaphthalene (1-MN) targeting monoaromatic hydrocarbons (MAH) mainly benzene, toluene and xylene (BTX) in a 16-fold fixed-bed reactor system at 400 ^{0}C and 30 bar. The CoMo/Beta (38) at NaOH concentration of 0.2M produced the highest BTX yield of 30% at 85% conversion of 1-MN without deactivation for reaction time of 24h. All catalysts were characterized using different characterization techniques including XRD and BET.

Development of new mesoporous zeolitic materials for refining processes

Sathiyamoorthy Murugesan, [a, b] Omer Refa Koseoglu, [c] Robert P. Hodgkins, [d] Kuo-Wei Huang, [b] and Jean-Marie Basset [b] *

[a] Interdisciplinary Research Centre for Refinery and Advanced Chemicals, KFUPM, Dhahran, 31261, Saudi Arabia. Dr. Sathiyamoorthy Murugesan: s.murugesan@kfupm.edu.sa

[b] King Abdullah University of Science & Technology, KAUST Catalysis Center (KCC), 23955-6900 Thuwal, Saudi Arabia.

[c] Catalyst Center of Excellence R&D Division, Research & Development Center, Saudi Aramco, Dhahran-31311, Saudi Arabia.

[d] Advanced Materials, Catalyst Center of Excellence R&D Division, Research & Development Center, Saudi Aramco, Dhahran-31311, Saudi Arabia.

Abstract:
Catalysis by zeolites is a crucial domain in hydrocarbon transformations. Zeolites are use as catalyst component, but they are not the final form of the catalysts. Predicting the reaction performance of activity and selectivity of zeolites from their properties is still a major challenge in the field. The main difficulty is the number and diversity of active sites, some of which are catalytically active. In particular, among those active sites, the acid strengths and redox properties vary dramatically depending on their structure and localisation on the surface. In our approach, well-defined zeolites are designed in which most of, if not all, the sites are structurally identical. It is known that treatment of fresh hierarchical mesoporous FH-ZSM-5 (**1**) at 700 °C led to mainly isolated silanols and Brønsted acid sites on the surface of FH-ZSM-5_{700} (**2**). The design of the Lewis and Brønsted acids is based on the drastic change of the framework of **2**. This framework was post-synthetically modified by substituting framework oxygen of mesoporous zeolite for nitrogen. This is achieved by a treatment of zeolites under ammonia flow at 500 °C. The resulting material contained two types of well-defined amine functionalities (**3**). In this same reaction, upon increasing temperature to 700 °C, mainly silazane bridges are formed in the framework of MFI-type (**4**). The functionalized well-defined ZSM-5 materials such as **2**, **3** and **4** would be subjected to surface organometallic chemistry and its applications, which are in a new avenue in surface science. Herein, we show the design of novel materials such as [L_nX_nM-O-ZSM-5, M = Ti, Zr, Hf, Ni, Pt], [L_nX_nM-NH-ZSM-5, M = Ti, Zr, Hf, Pt], [MNPs@ZSM-5, M = Ni, Pt] and MNPs@NH_2-ZSM-5, M = Ni, Pt] for refining processes.

Keywords: H-ZSM-5, Dehydroxylation, Surface functionalization, SOMC, Metal Nanoparticles.

Thermal liquefaction of waste polystyrene plastic to liquid hydrocarbon fuel using ethanol as a solvent

Nabeel Ahmad
Center for Refining and Advanced Chemicals, King Fahd University of Petroleum and Minerals, Dhahran 31261, Saudi Arabia
(E-mail: nabeel.ahmad@kfupm.edu.sa)

ABSTRACT

Polystyrene plastic was converted to hydrocarbon fuel via ethanolysis process under different conditions. Oil obtained was characterized using different analytical equipment. The highest yield of oil obtained was 84.7 wt% at 350 °C, C_2H_5OH/PS ratio 0.5:1, and 60 mins. The viscosity, density, pH, calorific value and flash point of the oil product at this condition was 0.36 cP, 0.88 g/mL, 6.86, 40.91 MJ/kg and 55 °C respectively. GC-MS analysis showed that the oil was mostly composed of aromatics, alkenes and alkyls; which made the liquid oil suitable to be used as fuel.

Oxidative Dehydrogenation of n-Butane to Produce Key Petrochemical Feedstocks

Ahmad Sabban,[1,2] Ziyauddin S. Qureshi,[1*] Gazali Tanimu,[1]
[1]Center for Refining & Advanced Chemicals, King Fahd University of Petroleum & Minerals, Dhahran 31261, Saudi Arabia.
[2]Research and Development Center, Saudi Arabian Oil Company, Dhahran 31311, Saudi Arabia.
E-mail: g201264800@kfupm.edu.sa; zqureshi@kfupm.edu.sa

Abstract

Oxidative dehydrogenation of n-butane reaction targeted at producing the highly demanded petrochemical feedstock's (ethylene, propylene, and butadiene) was investigated. Different Ni-Bi-O/ZSM-5 and WO_x/SBA-15: (i) 10 wt% Ni-10 wt% Bi on ZSM-5, (ii) 15 wt% Ni-10 wt% Bi on ZSM-5, (iii) 20 wt% Ni-10 wt% Bi on ZSM-5, (iv) 20 wt% Ni-30 wt% Bi on ZSM5, (v) 6 wt% tungsten on SBA-15 catalysts were prepared by wet co-impregnation method. The synthesized catalysts were characterized and tested for oxidative dehydrogenation of n-butane to produce light olefins. Different ratios of $Bi_{12}SiO_{20}$ and Bi_2SiO_5 phases were obtained depending on the Ni and Bi weight % loading. Five individual runs using Ni-Bi-O/ZSM-5 catalysts were carried out. Furthermore, ODH reaction was carried out using a dual bed catalyst system utilizing both Ni-Bi-O/ZSM-5 (dehydrogenation) and WO_x/SBA-15 (olefin metathesis) catalysts to enhance the propylene selectivity. In the dual bed catalyst system, two methods were utilized: (i) loading dehydrogenation and olefin metathesis catalysts one above the other, and (ii) mixed bed, which used both types of catalyst in equal amounts. Among individual catalysts, 15Ni-10Bi-O/ZSM-5 catalyst showed high activity and selectivity towards light olefins. Further combination of 15Ni-10Bi-O/ZSM-5 with 6WO_x/SBA-15 in dual bed system depicted the best performance with total selectivity of 13.7, 6.1, and 32.5 wt% for ethylene, propylene, and butadiene, respectively.

Keywords: Dehydrogenation, ZSM-5, butane, ethylene, propylene, butadiene

Controlled Autoxidation of Heavy Petroleum to Produce Carbon Fiber Precursors

Lahmady S. Mohamed, Mohammad M. Hossain, Muhammad N. Siddiquee

Department of Chemical Engineering, KFUPM
Interdisciplinary Research Center for Refining & Advanced Chemicals (IRC-RAC

Abstract

✓ Carbon fiber (CF) is ultra-light strong material that is useful in different industries like aviation, automobiles, military, and sports.

✓ There is a growing demand of CF, but traditional polyacrylonitrile (PAN) based carbon fiber is expensive that limits the extensive production and use of CF.

✓ Based on the cost and availability, heavy petroleum is promising to produce CF which is one of the few, higher value, and large global market products.

✓ This study highlights the prospects of carbon fiber precursor production from the heavy petroleum fractions following a cheap and sustainable technique .

Characterization Of Soot from the Atmospheric Combustion of Diesel/ Gasoline -Oxygenate Blends

Mohammed Ameen Ahmed Qasem [a], , Eid Al Mutairi [a], Abdul Gani Abdul Jameel [a,b*]

[a] *Department of Chemical Engineering, KFUPM, Dhahran, Saudi Arabia*

[b] *Center for Refining & Advanced Chemicals,KFUPM, Dhahran, Saudi Arabia*

Abstract

The soot formed during the diesel/gasoline -ethers blends combustion was characterized by a CHNS analyzer, spectroscopic techniques, and X-ray diffraction (XRD). The ethers used in this research were diethyl ether, dibutyl ether, dioctyl ether, tert-butyl-methyl ether (TBME), and diphenyl ether. For example, diethyl ether was added to diesel at 5 and 20%. The soot was collected from the smoke point lamp using Aluminum foil. The CHNS analyzer analysis of the Diesel/diethyl ether blend's soot showed that the soot particles were mostly composed of carbon with small percentages of nitrogen and sulfur. The EDX analysis showed that the blend soot also primally consists of carbon as well as oxygen. The carbon nanomaterials formed from the soot of diesel/gasoline -ethers blends appeared a non-uniform morphology. However, there are various grains are looked like carbon nanotubes and nanospheres at 1 and 0.50 μm as shown by JEOL SEM. The High-resolution transmission electron microscope, (HR-TEM) showed chain-like agglomerations at 100 nm. The shapes of individual particles of the nanomaterials of blend soot are spherical-like at 100 and 200 nm. The size average of the soot particles ranges from 40 to around 50 nm. The Raman spectra showed two major bands at 1590 cm-1 (G)and 1583 cm-1(D), indicating that the graphite-like structure was formed. The TGA analysis of the blend soot shows thermal stability. The thermogravimetric curve shows a sharp (at 600°C) and gradual (from 50°C to about 1000°C) weight loss under air and N_2 environment, respectively. Although the XRD pattern of the blends soot corresponds to the graphite of multi-walled carbon nanotubes, it generally shows amorphous material in association with nanotubes.

Turning Municipal Solid waste into Fuel via Integrated Thermochemical and Electrochemical Processes

Achmad F. P. Putra[a], Muflih A. Adnan[c,d], Mohammad M. Hossain[a,b], Md Golam Kibria[d*]

[a]Department of Chemical Engineering, [b]Center of Research Refining and Advanced Chemicals, King Fahd University of Petroleum & Minerals, Saudi Arabia

[c]Department of Chemical Engineering, Islamic University of Indonesia, Indonesia
[d]Department of Chemical & Petroleum Engineering, University of Calgary, AB, Canada

Abstract

Municipal solid waste (MSW) is a potential sustainable energy source due to its abundant availability and high energy content. Gasification is a thermochemical conversion process which enables converting solid feedstocks (i.e., MSW, biomass and coal) into gaseous products. It allows the conversion of toxic organic volatiles (tar) into useful syngas, which is difficult in direct waste oxidation/combustion processed. The present research investigated a new integrated gasification process to ensure complete decomposition of tar by splitting char and gas products. The produced char is directed to char combustor to produce electricity for powering CO_2-to-methanol electrolysis, while the gas products are sent to reforming section to decompose tar into syngas. A modified approach in thermodynamic simulation is used to estimate the performance of the proposed process. In contrast to the previous approaches which assumed only gases as a pyrolysis product, the present approach considers actual char composition in the biomass decomposition product. It is found that char composition is sensitive towards the syngas yield and the gasification temperature. The increase of char combustion from 20% to 50% promotes methanol production via electrolysis, but suppresses the overall energy productivity due to energy loss in the power generation and electrolysis. The optimum gasification performance is observed at the air/O_2 equivalence ratio of ~0.3 with the energy production efficiency of 80% and 82% for air and oxygen gasifying agents, respectively. These findings are consistent with the experimental counterpart as reported in the literature.
Keywords: Thermodynamic, gasification, municipal solid waste, simulation, electrolysis

CO_2 Assisted Oxidative Dehydrogenation of Propane to Propylene over MoO_3 based Catalyst

Suleiman Magaji[a], Majid L. Balogun [a], Sagir Adamu [a], Idris A. Bakare [b], Zuhair O. Malaibari[a,b], Mohammed S. Ba-Shammakh [a], Mohammad M. Hossain [a,b]

[a] Department of Chemical Engineering, King Fahd University of Petroleum and Minerals, Dhahran, Saudi Arabia.

[b] Center of Hydrogen and Energy Storage, King Fahd University of Petroleum and Minerals, Dhahran, Saudi Arabia.

Abstract

This study investigated the catalytic performance of a novel fluidizable MoO_3 supported on La_2O_3-γAl_2O_3 composite for oxidative dehydrogenation (ODH) of propane to propylene with and without CO_2. The La_2O_3-γAl_2O_3 composites were synthesized by a coprecipitation method, while the active MoO_3 was deposited by an incipient wetness impregnation approach. The incorporation of La_2O_3 controlled both the acidity and reducibility of the catalysts, as revealed by NH_3-TPD and H_2-TPR analysis, respectively. The N_2 adsorption/desorption isotherms analysis indicated the mesoporous pore size distribution of the catalysts. The ODH of propane experiments were developed in a fluidized CREC Riser Simulator by feeding propane with or without CO_2. The performance of the synthesized catalysts was found to be a function of the La_2O_3 content of the support. In general, the propane conversion declined slightly, while propylene selectivity increased with the increasing La_2O_3 content. It was hypothesized that the presence of basic La_2O_3 reduced the acidity and the non-selective sites of γ-Al_2O_3, which were responsible for deep oxidation of propylene and propane. Among the studied catalysts, MoO_3/La_2O_3-$\gamma Al_2O_3$1:2 displayed the highest propylene yield of 28.2 % at 600 °C in the absence of CO_2. However, a superior yield of 35.2% was achieved over the same catalyst by the introduction of CO_2-admixture in the feed at 550 °C.

Key words: Oxidative dehydrogenation; propane; propylene; mild oxidant; CO_2

Effects of metal support interaction on dry reforming of methane over Ni/Ce-Al_2O_3 catalysts

Ariel H. Gursida[a], Sagir Adamu[a], Abdul-Rashid Bawah[a], Oki Muraza[a], Zuhair Malaibari[a,b], Mohammad M. Hossain[a,b]

[a]Department of Chemical Engineering, Center for Refining and Advanced Chemicals, King Fahd University of Petroleum and Minerals Saudi Arabia

Abstract

CO_2 reforming (dry reforming) of methane proffers the most desirable alternative for the conversion of large amounts of methane and CO_2 to syngas. Methane constitutes a significant portion of natural gas (about 90%) and biogas; while shale gas contains about 27.5 – 85% CH_4, with up to 9 v/v % CO_2. This CH_4-rich shale gas is mainly consumed by the emerging natural gas power plants, which have long term potential usage as they progressively supplement the coal power plants. On the other hand, about 32 gigatonnes of CO_2 is produced annually from the combustion of fossil fuels. Therefore, dry reforming of methane will help to recycle CO_2 by utilizing a fraction of the pool of methane available from the aforementioned sources. DRM offers the following advantages: (1) DRM utilizes the two major greenhouse gases (i.e. CO_2 and CH_4) as feedstock; (2) DRM can produce syngas with a desirable H_2/CO ratio, which may be used in Fischer–Tropsch process, or fuel cell applications among others; (3) utilization of natural gas, biogas and shale gas containing significant quantities of CH_4; (4) the endothermic nature of the reaction between CO_2 and CH_4 makes it a suitable candidate for chemical energy transmission systems.

In this study dry (CO_2) reforming of methane is conducted over two newly synthesized Ni20/Ce-γAl_2O_3 and Ni20/Ce-meso-Al_2O_3 catalysts. XRD patterns indicate that Ni20/Ce-meso-Al_2O_3 exhibits a better dispersion of nickel, while Ni20/Ce-γAl_2O_3 has larger amounts of nickel crystallites. TPD kinetics analysis suggests that Ni20/Ce-meso-Al_2O_3 has lesser metal-support interaction than the Ni20/Ce-γAl_2O_3. TGA indicates that incorporation of ceria into the Al_2O_3 matrix helps to stabilize Ni(20)/Ce-meso-Al_2O_3 during dry reforming of methane. TPR indicates that the synthesized catalysts are sufficiently reducible below 750 °C. A fixed bed reactor evaluation (at 750 °C) shows that both catalysts can facilitate methane reforming to syngas with minimal coking throughout the 30 h TOS. However, Ni20/Ce-meso-Al_2O_3 shows better prospects in terms of prolonged stability for dry reforming applications. Moreover, the syngas yield for Ni20/Ce-γAl_2O_3 is very close to equilibrium prediction during the first one (1) hour reaction time.

Key words: Ni/Ce-Al_2O_3; metal-support interaction, Dry reforming; H_2 yield; syngas.